T0123147

# Defending the Nation

## U.S. Policymaking to Create Scientists and Engineers from Sputnik to the 'War against Terrorism'

Juan C. Lucena

UNIVERSITY PRESS OF AMERICA,® INC.
*Lanham • Boulder • New York • Toronto • Oxford*

4501 Forbes Boulevard
Suite 200
Lanham, Maryland 20706
UPA Acquisitions Department (301) 459-3366

PO Box 317
Oxford
OX2 9RU, UK


Printed in the United States of America
British Library Cataloging in Publication Information Available

Library of Congress Control Number: 2005924238
ISBN 0-7618-3156-8 (clothbound : alk. ppr.)
ISBN 0-7618-3157-6 (paperback : alk. ppr.)

∞™ The paper used in this publication meets the minimum requirements of American National Standard for Information Sciences—Permanence of Paper for Printed Library Materials, ANSI Z39.48—1992

# Contents

# Preface and Acknowledgements

My first encounter with the relationship between the American nation and technoscientific education took place during my engineering undergraduate years at Rensselaer Polytechnic Institute (RPI) in the mid-1980s. Throughout my engineering courses, I encountered hundreds of problems about fluid flow, rocket propulsion, and 'dog fight' flight patterns of U.S. war planes chasing Soviet MIGs, among others, all clearly assuming the Cold War as a context of application. Meanwhile, my design courses began to emphasize manufacturing processes and the use of new materials. At that time, I did not understand that my engineering-science faculty remained challenged by the needs of the Cold War, while my design faculty was beginning to be challenged by the successes of Japanese manufacturing. Neither I nor my engineering faculty could see that our decisions to learn and teach these kinds of problems and designs were responses to historically and culturally specific challenges.

After finishing my engineering degrees in mechanical and aeronautical engineering (1987–88) and wondering about the meaning of the Cold War's end to the future of engineering education and practice, I became a graduate student in Science and Technology Studies (STS) at RPI. Under the wonderful guidance of STS faculty Deborah Johnson, Sal Restivo, Langdon Winner, and Rick Worthington, I conducted ethnographic research in a center for composite materials where I witnessed how, under projects funded by the Pentagon and companies like Boeing and Alcoa, scientists and engineers responded to the needs of both the Cold War and industrial competitiveness. I would like to thank these committed professors for their guidance and teaching of STS concepts and methods that allowed me for the first time to understand the relationship between national needs and technoscientific research and education.

After two years of STS graduate education at RPI, I worked at the National Science Foundation (NSF) where I witnessed firsthand how policymakers in science and engineering research and education struggled with the challenges posed by both the end of the Cold War and the successes of Japanese technology. As noted in chapter 4, I participated directly in the dissemination of the science and engineering pipeline, collecting demographic data and creating presentations for NSF officials on the state of science and engineering education. Eventually I came to manage a small program to fund women and minorities to go into graduate engineering education. I would like to thank Rachelle Hollander, then program manager for Ethics and Values in Science and Engineering (EVIST), for her support and mentoring during this time at NSF.

Wanting to further theorize and research past, present, and future challenges to science and engineering education, I enrolled in the Ph.D. program in STS at Virginia Tech. There Gary Downey gave me the opportunity to participate in an ethnographic research project where we observed how problem solving in the engineering sciences, privileged in a curriculum that is the legacy of the Cold War, challenged engineering students as persons. Under the guidance of a wonderful committee, I completed my Ph.D. dissertation that served as background for this book. To Gary, my friend, colleague, and mentor, I am forever indebted for his unconditional support during the most challenging times of both my personal life and academic career. He directed my first steps in both teaching and research and taught me tolerance of human differences. To Richard Hirsh, I am grateful for his friendship, humor, and for teaching me how to write concisely, clearly, and to the point for the unforgiving world of academic publishing. To Skip Fuhrman, I am thankful for not letting me forget the value of Marx, Weber, and Durkheim during these postmodern times. He gave me the gift not only of sociological theories but also of his own personal experiences which helped me during difficult times of soul-searching. Through the example of his career, Steve Fuller taught me the meaning of networking and activism from within. Not only did he challenge me during both doctoral examinations but gave me invaluable lessons of the value of STS as a tool for social and political change. With Langdon Winner, I learned about the possibilities of a more democratic political life through critical participation in technological designs and practices. His commitment for social justice has always been an invaluable example. To all of them, I extend my most sincere gratitude.

During my first faculty appointment, I received the unconditional support of my then department chair Peter Quigley to research and write under an NSF-CAREER award (SDEST-9875089). This time and support allowed me to interview dozens of practitioners and educators and begin conceptualizing the impact of globalization on engineering education and practice. This research confirmed not only the strong relationship between U.S. national needs and technoscientific education and practice but how this relationship differs across national contexts.

Now at the Colorado School of Mines (CSM) where I research and teach on the differences in engineering education, knowledge, and practice among nations, I have encountered great support from colleagues at the Division of Liberal Arts and International Studies (LAIS). Particularly, I want to thank Arthur Sacks for his continuous wise and timely mentoring, Carl Mitcham for his collegiality and relentless encouragement to finish this book, and Jon Leydens for his friendship and lessons to teach my students to become better writers.

Throughout this scholarly journey, that began at the eve of the end of the Cold War and pauses now in the midst of the so-called 'war against terrorism', my Colombian and U.S. families have always been there to support me. To Liz, *mi alma gemela*, I am grateful for giving me the gift of true companionship as well as endless hours of editing. She alone put up with my anxieties and fears as I wrote my dissertation and this book. She alone had to endure my clashes with the English language. To her, all my love and admiration. To my children Juju and Nico, I am thankful for pulling me away every morning from this manuscript to eat breakfast, brush their teeth, finish homework, and walk to their bus stop. Hopefully, when they grow up and have the choice to study science or engineering, U.S. policymakers and educators will be challenged with building a kinder, socially progressive, and environmentally cleaner world. To Juju and Nico, I dedicate this book.

I gratefully acknowledge permission to reprint the following. An earlier version of chapter 1 appeared as "From Frontier to Terrorism: Toward an Interdisciplinary Assessment of Science Education Policy Making," *Philosophy Today*, Supplement (2004), pp. 56–64, reprinted by permission of DePaul University. Earlier text of chapters 3 and 4 appeared in "Making Women and Minorities in Science and Engineering: Nation, NSF, and Policy for Statistical Categories," *Journal of Women and Minorities in Science and Engineering*, 6 (1), pp. 1–32, reprinted by permission of Begell House, and in "Women in Engineering: History and Politics of a Struggle," *IEEE Technology and Society Magazine*, (Winter), pp. 185–194, reprinted by permission of IEEE. Modified parts of chapter 5 appeared in "Flexible Engineers: History, challenges, and opportunities for engineering education," *Bulletin of Science, Technology, and Society*, 23 (6), pp. 419–435, reprinted by permission of SAGE Publications. Special thanks to my friend Luzia Ornelas for her artistic talent in the design of the book cover.

# *Chapter One*

# Introduction

"I'm proud of you. We are depending on you to develop the tools we need to lift the dark threat of terrorism for our nation—and for that matter, the world. All of us here today, whether we're scientists or engineers or elected officials, share a great calling. It's an honor to participate in a noble cause that's larger than ourselves." President George W. Bush, "Anti-Terrorism Technology Key to Homeland Security", speech delivered to scientists and engineers at the Argonne National Laboratory on July 2002.

President Bush's call for scientists and engineers to save the nation from terrorism was not the only one, nor even the first. Just four months after the tragic events of September 11, 2001, Rita Colwell, then Director of the National Science Foundation (NSF), delivered a speech entitled "Science as Patriotism" to scientists and engineers at the annual meeting of the Universities Research Association in which she said:

"Every discussion, whether it is about airline safety, emerging diseases, failure of communication links, bioterrorism directed at our food and drinking water, assessment of damaged infrastructure, the mind/body response trauma, or a myriad of other concerns, depends on our scientific and technical knowledge. . . . We need to make a commitment to a home-grown science and engineering workforce that uses the diversity of our national as the talent pool. If the science community can be hands-on to inspire young people to a future in science, we would be performing one of the most enduring acts of patriotism for the nation" (Colwell 2002, p. 1).

Soon afterwards, the National Research Council's (NRC) Committee on Science and Technology for Countering Terrorism issued *Making the Nation Safer: The Role of Science and Technology in Countering Terrorism* in which the Committee acknowledged that "America's historical strength in science and engineering is perhaps its most critical asset in countering terrorism without degrading our quality of life. . . . *The nation's ability to perform the needed short- and long-term research and development rests fundamentally on a strong scientific and engineering workforce*. Here there is cause for concern, as the number of American students interested in science and engineering careers is declining . . ." (National Research Council 2002, p. 23) (italics mine). Immediately afterward, scientists at NSF's Directorate for Mathematical and Physical Sciences (MPS) heeded these calls for a response to terrorism and by 2003 issued a program solicitation entitled *Approaches to Combat Terrorism (ACT): Opportunities in Basic Research in the Mathematical and Physical Sciences with the Potential to Contribute to National Security*, with an expected initial funding of $3.5 million. By June 2003, NSF had funded $20 million worth of exploratory research and education programs dealing with terrorism.

## WHAT IS GOING ON HERE?

The emergence of an image of the American nation under the threat of terrorism has allowed the President, the NSF Director and others to issue compelling calls to scientists and engineers to save the nation. At the same time, scientists and engineers working in and out of the federal government, such as those involved in advisory committees for NSF and the NRC, have appropriated this image and these calls to create new programs and to secure federal monies for education, research, and workforce programs in order to save the American nation from terrorism. Scientists, engineers, and administrators in higher education are already applying for grants to develop projects in recruitment, curriculum, and infrastructure. Already, many colleges and universities have developed programs to train students for jobs in homeland defense (Barlett 2003). Many students have responded to the image, the calls, and the programmatic opportunities by enrolling in science and engineering programs. This entire process, from the emergence of images of the American nation under threat, to the creation of a discourse of saving the nation with science and technology, to the development of federally-funded programs to educate scientists and engineers, is what I call *policymaking to create scientists and engineers.*

However, this process of policymaking is nothing new. What has changed are the image of the American nation under threat, the actors making the calls, the content of their rhetoric, and the characteristics desired in scientists and engineers in order to fight a new threat. But the policymaking *process*—from the emergence of an image of a nation under threat and its associated discourse to the struggle of actors for budget allocations and the creation of programs—has remained strikingly consistent in the last five decades of U.S. history. Since the end of World War II, scientists and engineers have often been called to save the American nation from an external threat. In the 1960s the calls, programs and monies went for the creation of a top cadre of *elite scientists* to develop the basic science that would save the American nation from the threat of communism, particularly after the USSR had successfully launched Sputnik in 1957. In the 1970s the calls went out for *appropriate scientists and engineers* who would solve the nation's domestic and environmental problems that it had faced throughout that decade. In the 1980s, *large numbers of engineers* were called to address the competitive Japanese economic challenge to the American nation. Since the early 1990s *flexible scientists and engineers* have been called to research and solve problems in an increasingly competitive global economy. More recently, *patriotic scientists and engineers* are being called to develop the knowledge and infrastructure to keep America safe from terrorism. This book is a history, from Sputnik to the first years of the 21st century, of these transformations in policymaking to create scientists and engineers in the United States.

## WHY IS THIS HISTORY IMPORTANT?

Harvey Averch has argued that policymakers resort to a diverse range of rhetorical strategies in order to justify federally-funded programs. They use a wealth of metaphors, information, models, witnesses, coalition building, etc. in order to create these strategies that would, they hope, lead to the development and implementation of policies and/or securing of federal monies for programs to educate scientists and engineers. Averch concludes that at the federal level, science and technology policy is limited to a few key issues: money, priorities, and the parties that gets them (Averch 1985). I argue here that policymaking is about more than money. It is about defining and solving problems for a nation under threat. This book goes beyond identifying rhetorical strategies and models to placing them in a larger historical and cultural context. For example, this book explains the rise and demise of the science and engineering pipeline model, how this model became a powerful metaphor within the rhetorical strategies deployed by those demanding more federal dollars for education and workforce programs, and how faculty and administrators in colleges and universities capitalized on the pipeline. But more importantly, it shows how these strategies, in order to gain legitimacy, required a dominant image of the American nation under threat by Japanese technological competitiveness.

Just about every organization related to science and engineering has a unit or program that deals with scientific and engineering workforce in which professional staffs and executives intervene in policymaking by developing knowledge and implementing programs for scientists and engineers. Examples of these include the Office of Education and Human Resources of the American Association for Advancement of Science (AAAS), the Engineering Workforce Commission of the American Association of Engineering Societies (AAES), the National Action Council for Minorities in Engineering (NACME), the Division of Science Resource Statistics at the National Science Foundation (NSF), and the Committee on Diversity in the Engineering Workforce of the National Academy of Engineering (NAE). At different points in their histories, actors in these organizations have appropriated, used, and sometimes transformed the calls and models that emerge under images of the American nation under threat. Many professionals have made careers out of developing models or generating reports, appropriate to the times, that make the next best case for particular types of programs to educate scientists and engineers. But these actions do not occur in a vacuum. The funding and attention that they receive depend heavily on how aligned their arguments are with the image of the American nation under threat. This book provides a deeper understanding of their activities, and their successes and failures, by placing them in a larger historical and cultural context.

The federal programs funding these activities, usually located at the NSF, also play a part in this process of policymaking for scientists and engineers. This book helps academic administrators and faculty who recruit, retain, and educate scientists and engineers understand the larger cultural and historical dimensions of the changes in federal priorities and budgets, and the consequent emergence and demise of programs at colleges and universities. For example, during the 1995 national conference of the National Association of Minority Engineering Program Administrators (NAMEPA) under the theme "Partners in the Pipeline" my early insights on the pipeline, now contained in this book, allowed key officers in the organization to understand that their effort was culturally and historically located in a time when the American nation needed large numbers of engineers to respond to the Japanese challenge. NAMEPA's officers used this understanding to build successful strategies to secure more visibility, monies, and access to power.

In two ways, this book also contributes to our understanding of the history of policymaking with regards to science and technology. First, it introduces a new concept of culture as images of a nation under threat, which challenge different actors to struggle to define problems and solutions in terms of policies and programs to educate and create scientists and engineers for specific national needs. This theoretical connection between culture and policy is gaining increasing attention, as evidenced for example by the increasing number of conference sessions and papers that attempted to draw such connections at recent meetings of the American Political Science Association (APSA) and the International Studies Association (ISA). Second, this book calls attention to the making of scientists and engineers as an often ignored but important problem in governing the nation-state. Due to the increasing importance of science and technology in all areas of American life, the making of scientists and engineers has become a significant governmental act to ensure national security, economic growth, and, sometimes, social well-being.

## THEORETICAL FRAMEWORK

As are all histories, this one is informed by theories: first, by a cultural theory that explains how images of nation challenge all the actors involved in policymaking; second, by a theory of governmentality that explains the making of scientists and engineers as a problem for the sustenance of the nation-state; third, by a theory of discourse explaining why certain things can be said, how they can be said, and who can say them; and fourth by a theory of social construction that helps us understand the processes that lead to the development of reports, models, statistics, etc. about scientists and engineers.

### Culture as Images of Nation

Some cultural theorists have begun calling for new models of culture. In light of the large and fast mobility of peoples, ideas, customs, and beliefs around the world, they have questioned the usefulness of traditional models of culture in which members of a bounded community share a common set of values, beliefs, norms, and behaviors. For example, George Marcus calls for a new way to talk about the relationship between culture and individuals. "The languages that have been useful in talking about culture and politics in the past don't really seem adequate to this historical moment" (Marcus 1999, p. 67). Gary Downey has proposed a new concept of culture in which individuals living and working in a particular spatial and temporal location are challenged by dominant images. Dominant images create expectations about how individuals in that location are supposed to act or behave. In this new concept of culture, the image remains the same over a period of time, while individual or group reactions to the image's challenges might differ. When challenged by the same image, individuals or groups resist, accommodate, fully accept, or experience ambiguity in different ways (Downey 1998; Downey and Lucena 2004). For example, after September 11, 2001, a dominant image emerged of the American nation under the threat of terrorism. All Americans are now challenged by this image. Some have responded by enlisting in the armed forces while others have enrolled in academic programs dealing with international and homeland security. Others have engaged in patriotic behavior at home, other are busy writing research grants on biometrics, while others have resisted the actions of government. In sum, the challenge by the new dominant image of America is the same for all Americans, while the reactions differ.

Although individual experiences might be different, we can recognize patterns among experiences. For example, when President Bush and NSF Director Rita Colwell reacted similarly to the emerging image of the American nation under the treat of terrorism by calling scientists and engineers to be patriotic, we could begin to observe a pattern of behavior as other members in the policymaking process reacted in similar ways, as for example the scientists and engineers in the advisory committees for NRC and NSF have done. There might be others who have resisted the challenges of this image and argued, for example, that enlisting scientists and engineers in the war against terrorism might prove counter-productive to scientific creativity and technological innovation. As we will see, those who embrace the challenges of the dominant image of the American nation under threat have an easier time getting what they want.

This book is a history of the emerging dominant images of a nation under threat, of the challenges that these images pose to policymaking groups, and of the subsequent reactions, as manifested in the actions of legislators and policymakers to create programs to educate scientists and engineers in particular ways.

## Scientists, Engineers and the Nation-State

According to French philosopher Michel Foucault, the central problem of governance since the 19th century became managing the population of the nation-state. This problem led governments to implement techniques of power to reconfigure individuals into specific categories that would allow the nation-state to survive. As Foucault argues, these techniques of power "were never more important or more valorized than at the moment when it became important to manage a population: the managing of a population not only concerns the collective mass of phenomena, the level of its aggregate effects, it also implies the management of population in its depths and details" (Foucault [1978]1991, p. 102). Furthermore, he argues that "[p]opulation is the object that government must take into account in all its observations and *savoir*, in order to be able to govern effectively in a rational and conscious manner. The constitution of a *savoir* of government is absolutely inseparable from that of a knowledge of all the processes related to population in its larger sense: that is to say, what we know call the economy" (Ibid., p. 100).

Contemporary foucauldian theorists argue that these techniques of power, or "technologies of government," as they also call them, have become instrumental in managing population and hence in the survival of the modern nation-state. Miller and Rose understand these "technologies of government", as "mechanisms of objectification of individuals through which authorities of various sorts have sought to shape, normalize and instrumentalize the conduct, thought, decisions and aspirations of others in order to achieve the objectives they consider desirable" (Miller and Rose 1993). In one of his last writings, Foucault himself referred to the coordinated ensemble of these technologies of government as "governmentality." This ensemble "which has as its target population, as its principal form of knowledge political economy, and as its essential technical means apparatuses of security . . . is what has permitted the [modern] state to survive" (Foucault [1978]1991, p. 103).

The study of governmentality has become a particular *ethos* of analysis "marked by a desire to analyze contemporary political rationalities as technical embodiments of mentalities for the government of conduct" (Miller and Rose 1993, p. 76). Specifically, these analyses focus on new ways to analyze the exercise of political power in advanced liberal democracies by means of technologies of government that eventually lead to the "shaping of the private self."[1] Ultimately, these technologies of government are intended to bring social and economic order in contemporary liberal governments by serving both ends of economic government: political economy and social security (Gordon 1991). The former is served through the efficient allocation of population in the different sectors of the economy, for example, by counting, identifying, and predicting the number of engineers needed in key industries or scientists in areas of strategic importance for national security. Social security is served through the creation of self-regulated social citizens who will eventually integrate themselves into the economy, for example, by providing educational opportunities to individuals who can eventually join the workforce. Situating and tracing power and knowledge that ultimately allow the deployment of technologies of government requires an attention to *language*. In this book, I analyze the emergence and demise of technologies of government (e.g., supply/demand workforce models, supply-side models such as the pipeline, etc.) as they relate to the dominant image of nation that challenges the creators of these technologies.

## Policymaking as a Discursive Field

As Miller and Rose have recognized, "[g]overnmentality has a discursive character: to analyze the conceptualizations, explanations and calculations that inhabit the governmental field requires an attention to language" (Miller and Rose 1993, p.78–9). This is why I focus my attention on the language of actors (government officials, representatives of industry, advocates of underrepresented groups in science and engineering) as they struggle to define national problems and their solutions in terms of scientists and engineers. However, as Miller and Rose further suggest, in order to understand policy, it must be situated within a discursive field that gives legitimacy to its means and ends: "we suggest that policy should be located within a wider discursive field in which conceptions of the proper ends and means of government are articulated" (Ibid.). This is why I locate the language of policymaking either in official statements, in ad hoc reports, or in written policy within the context of time-specific images of nation. As language resonates with the image of the nation under threat, it acquires a legitimacy that allows policymakers to define problems and solutions in terms of scientists and engineers.

Miller and Rose do not provide a clear vision of where this discourse is located and how to access it. A clue comes from Foucault's writings on politics and the study of discourse, in which he points at *archaeology* as the methodology and the *archive* as the location where one finds a discourse. In studying discourse, he writes, "I am not doing exegesis, but an *archaeology*, that is to say, as its name indicates only too obviously, the description of an *archive*. By this word, I do not mean the mass of texts gathered together at a given period . . . I mean the *set of rules* which at a given period and for a given society define . . . the limits and forms of the *sayable*. What is it possible to speak of? What is the constituted domain of discourse? . . . Which utterances does everyone recognize as valid, or debatable, or definitely invalid? Which has been abandoned as negligible, and which has been excluded as foreign? . . . The limits and forms of *appropriation*. What individuals, what groups or classes have access to a particular kind of discourse? How is the relationship institutionalized between the discourse, speakers and its destined audience? . . . How is the struggle for control of discourses conducted between classes, nations, cultural or ethnic collectivities?" (Foucault [1968]1991, p. 58–60).

After having worked with NSF's programs in education and human resources in engineering, and after having read numerous reports calling for scientists and engineers to solve many of America's problems, I discerned a particular discourse, or set of rules, that gave legitimacy to everything NSF administrators and policymakers said, proposed, and did about educating and training scientists and engineers in the U.S.. As I researched the archives covering the last five decades in which actors and groups proposed actions for the education and training of scientists and engineers, I noticed the emergence of a different discourse for each decade. For example, in the 1960's we see a discourse of *elite science* that sets the "limits of the sayable and appropriation" to deal with the challenges posed by an image of the American nation under threat by Soviet science. In the 1970's we see a discourse of *appropriate science and technology* to deal with the challenges of an image of the American nation threatened by social and environmental problems. In the 1980s, under the dominant image of America under the threat of Japanese technological competition, *technology for economic competitiveness* emerges as a discourse. In the 1990s, *flexible technoscience for global competition* emerges as a discourse under the challenges of an image of America threatened by global competitiveness. And nowadays, we are witnessing the emergence of a discourse of *patriotic science and technology* to deal with the challenges posed by an image of America under the threat of terrorism. For each decade, these discourses have defined the *limits of the sayable* and the *limits of appropriation* about educating and training scientists and engineers. More than money and politics, images and discourse have shaped the actions of government to educate scientists and engineers in particular ways.

## The Construction of Knowledge about Scientists and Engineers

Locating the exact origin or end of a discourse is not that important. What is important in analyzing discourse, as Foucault argues, "is the law of existence of statements, that which rendered them possible—them and none other in their place; the conditions of their singular emergence; the correlation with other previous or simultaneous events, discoursive or otherwise" (Foucault [1968]1991, p. 59–60). Hence, my interest in analyzing statements made in the media is to see how the discourse about the nation travels across different areas of national life, for example, from public domains into official domains and vice versa. I am particularly interested in the "law of existence" of official statements about the needs for scientists and engineers: Who is permitted, or is able, to pronounce these statements? Who is excluded altogether from making statements? Which concepts of nation are valid, and which are not?

Foucauldian theorists have given us hints as to what might constitute a statement, let's say about population management, but they do not tell us how these become accepted as official knowledge through which the population is actually managed. As Miller and Rose claim, "discourse requires attention to particular technical devices of writing, listing, numbering, and computing that render a realm into a knowable, calculable and administrable object. 'Knowing' an object in such a way is more than a purely speculative activity: it requires the invention of procedures of notation, ways of collecting and presenting statistics. . . It is through such procedures of inscription that the diverse domains of 'governmentality' are made up, that objects such as the economy, the enterprise, the social field and the family [including science and technology] are rendered in a particular conceptual form and made amenable to intervention and regulation [policy]" (Miller and Rose 1993, p. 79). However, we should not assume that statements containing similar "procedures of inscription" (tables, statistics, graphs, models, etc.) have the same level of acceptance by legislators and policymakers. Actually, most statements about scientists and engineers during the 1980s contained sound statistics, tables, graphs, etc. to support their arguments, but not all were accepted by Congress as official knowledge to influence policy, even when the statements were aligned with the dominant image of the nation under threat. The questions regarding who writes such statements, who endorses them, what kinds of networks of economic and political power support them or attack them, have significant relevance to whether or not these statements finally become official knowledge. To analyze how statements become legitimate official knowledge, I turn to social constructivism.

Official knowledge about scientists and engineers emerges from a process of social construction in which individuals and groups with different interests pertinent to how scientists and engineers will be educated and trained compete and negotiate for solutions. Besides alignment with the dominant image of the nation and the emergent discourse, the success of statements in becoming official knowledge depends to a great extent on the how well the actors making the statements deploy their allies and resources. Nowhere is this more evident that in policy making. Interest groups such as universities and industry, and federal agencies, like NSF, deploy allies (congress persons, lobbyists, expert witnesses) and resources (statistics, reports, visual metaphors) to legitimate their claims about national problems and their possible solutions. But more than the result of negotiation or the appropriate deployment of allies and resources, official knowledge is the result of power struggles for the control of reality. Groups and individuals struggle to define national reality, first, by aligning themselves with the dominant image of nation and appropriating its emerging discourse, and, second, by defining the problems and solutions in their own terms. Successful groups and actors in this process will eventually shape programs, budgets and the meaning of how we understand the terms "scientist" and "engineer" in the U.S.

As a theory, social constructivism has been criticized for a number of conceptual limitations: first, for its neglect of the power differences between social groups or actors who are active in negotiating for knowledge and those silent voices who never make it into the process; second, for its neglect of structural relationships between classes, races, and genders; and, third, for its problematic conclusion that knowledge is a final product once the process of construction has reached closure or after actors appear to have settled a controversy.[2] My analysis of the construction of official knowledge for the making of scientists and engineers takes these limitations into consideration, for I do consider the power dimensions among those who participate and those who do not.

As we will see, the appropriateness of official statements does not depend solely on how well they resonate with the image of a nation under threat and its emergent discourse. For an official statement to influence policy, it takes more than addressing the nation's problems and proposing sound solutions. The power of official statements to influence policy also depends on who makes the statements, obviously within the limits of what is sayable. Groups and actors guarantee their participation in policymaking by their position relative to economic and political power. In the 1960s, for example, a scientific elite excluded the voices of working classes and racial minorities from policymaking that directed science to save the American nation from the threat of Soviet communism. But, more generally, even after a field of participants has been established, those who participate are not necessarily equal to one another. Among participants, significant power differences exist along race, gender, and class lines, and in the construction of official knowledge, these differences cannot be ignored. For example, in the 1980s when advocates for women and minorities helped create knowledge about scientists and engineers, they did not occupy the same position of power and influence as the Vice President of IBM, who later became the chairman of the NSB. Likewise, for actors and groups with different levels of power, policy outcomes have different implications. Throughout this book, I address the following questions: In science and engineering, whose knowledge claims are being voiced and whose are being silenced in the construction of official knowledge about education and human resources? Who benefits more from a policy that results from this knowledge? What purpose does this knowledge serve?

After analyzing the construction of official knowledge under different images and discourses of nation and the existing threat to nation, I look at how specific policies that emerge from this process aim at educating and training scientists and engineers.

As we can see, I have expanded the meaning of policymaking. It is more than just struggles among actors for money and power. Policymaking is a process that takes place under the challenges of dominant images and emergent discourses of a nation under threat. It is a struggle among actors and groups with different levels of power to shape knowledge about scientists and engineers and to influence the budgets and programs that locate, educate, train, and redirect them in new ways.

## THE NATIONAL SCIENCE FOUNDATION (NSF)

Nowhere is the cultural relationship between images of the nation and policymaking for scientists and engineers more evident than around the programs in education and human resources at the National Science Foundation (NSF). The NSF has emerged as the United States' leading voice in science and engineering issues, especially in educating and developing human resources for specific national needs. Given the increasing importance of scientists and engineers to fulfill national missions, such as the space race after Sputnik or the current war against terrorism, educating, training, and knowing about these issues have become important activities for federal involvement. The NSF has become the federal government's main instrument to help fulfill these responsibilities. The NSF's increasing influence in policymaking for education and human resources in science and engineering, particularly in the last 20 years, makes it a unique site in which to study how images of the American nation and its emerging discourses translate into budgets and programs to educate scientists and engineers. As new images of the American nation and discourses emerge, actors struggle to appropriate them and establish prescriptions of what and how the NSF should do to help save the nation.

As a source of certified knowledge about U.S. scientists and engineers, the NSF has gone from playing a reactive role to occupying the leading voice that informs the U.S. government about specific needs for scientists and engineers. Fifty years ago, as a young federal agency, NSF rarely pronounced public statements about the state of American science and technology. After the Soviet Union launched Sputnik in 1957, headlines describing a national crisis in science and education rarely relied on knowledge created at NSF.[3] Today, however, NSF has become the legitimate source of knowledge about scientists and engineers. "The New Global Workforce: High Tech Skills All Over the Map," "Shortage of Scientists Approaches a Crisis. . . ," "Wanted: 675,000 Future Scientists and Engineers," and "Scientists can help battle terrorism, NSF Director urges" are just few of the recent national headlines that rely primarily on knowledge made at NSF about the nation's needs for scientists and engineers to compete in a post-Cold War global economy and to protect the nation against terrorism.[4] In contrast with its previous, more passive role, NSF now takes the lead in informing the nation about how to educate and train its scientists and engineers to fit specific national needs.

### NSF as Technology of Government

Signed into law on May 10, 1950, by President Truman, the NSF Act of 1950 authorized and directed the new Foundation "[t]o promote the progress of science, to advance the national health, prosperity and welfare, to secure the national defense, and [to fulfill] other purposes" (Public Law 81-507). Although Congress did not intend for NSF to be a *mission-oriented* agency, like the Department of Agriculture or NASA, the Act of 1950 implicitly defined a national mission for NSF. Congress, NSF officials, and interest groups have used the Act of 1950 as the legislative mechanism to reinterpret the NSF's new missions throughout its 55-year history, redefining its research, education, and human-resource programs according to the nation's emerging needs. Today, for example, congressional committees invite groups to discuss how "new opportunities and challenges [which] have been created by the end of the cold war, the rise of multilateral economic competition from abroad, and the emergence of global environmental problems" might redirect the national mission of NSF (U.S. House Committee on Science 1993).

The legislative process by itself, however, is not enough to shape NSF's national mission when new national needs emerge. Within a cultural space defined by an image of the nation under threat, and the limits of what is sayable framed by the emergent discourse, groups struggle to redefine the national problems to be solved by NSF. As new images of the nation emerge, groups struggle to define the meaning of such terms as "national health," "national prosperity and welfare," and "national defense," and seek to propose solutions through NSF. These struggles have been particularly visible within NSF's education and human resource programs. In the last four decades, NSF has become the most important, and in some respects the only, federal agency in charge of developing and promoting science and engineering education and human resources.[5] From the launching of Sputnik in 1957 to the present, NSF has responded to different needs for scientists and engineers brought upon by national crises in two ways: its human-resource programs, such as the National Register and Manpower[6] Studies, and its education programs.[7] These programs are the mechanisms through which NSF helps supply and educate the scientists and engineers that the nation requires. For example, the National Register program, whose aim is to "make possible the location and identification of individuals with specialized skills when needed for Governmental purposes, including mobilization," has served to locate and identify scientists and engineers according to time-specific national needs. Similarly, the Manpower Studies activity, now the Division of Science Resource Statistics (SRS), as "the central program in the federal government for the provision of data on the supply, demand, education, and characteristics of the Nation's scientific and technical personnel resources," has produced projections for time-specific, supply-and-demand national scenarios. In the last 50 years, as new images of the American nation have emerged, these programs have changed their names, and even their location within NSF, but not their broad objectives: to locate and project scientists and engineers according to emerging national needs. In short, these programs have become technologies of government.

Locating, identifying, projecting, and hence (re)defining population categories of scientists and engineers is only one side of NSF's role in creating scientists and engineers. The other side is implementing science education programs aimed at producing the kinds of scientists and engineers that the nation needs to survive. Since NSF's early years, whether in the form of Fellowships, Teacher Training, or Curriculum Improvement, these programs have been aimed at the "development of the individual scientist" while ensuring the "produc[tion] of adequate numbers of young scientists and engineers qualified to do the things our national goals require"(National Science Foundation 1960).

NSF data and projections have become the most legitimate source of information regarding the state of science and engineering in the U.S. As NSF Historian, J. Merton England, reports, "[already] by [the] late 1950s more and more graphs, charts, and tables in books and articles carried the notation 'Source: National Science Foundation,' a designation that was becoming a stamp of authenticity. However shaky NSF's figures on scientific personnel . . . might be, and they were largely estimates, they were far more accurate than those available before and were becoming steadily better" (England 1982, p. 254). Nowadays, popular and academic media, educators, policymakers, etc. depend heavily on NSF's evaluations of the health of American science and engineering, particularly as it compares with that of other industrialized countries. An interesting outcome of NSF's information and projections is that it has guided policies and subsequent budget allocations to NSF programs, thereby creating a conflict of interest. Through these projections, NSF has informed, recommended, and shaped the policymaking process in science and engineering. Legislators, NSF officials, and other interested parties have been using NSF projections during appropriation and authorization hearings for NSF programs, legitimizing NSF as the source of knowledge that shapes its own policies. But more than just a bureaucracy endeavoring to perpetuate itself, NSF has emerged as a unique institutional solution to the political and economic problems surrounding human-resource development during the last 40 years. More than any other federal agency, NSF has become a federal instrument for allocating people in science and engineering fields without direct federal intervention or centralized policies of human-resource allocation, such as the policies followed in the former USSR. Constitutional hurdles and the lack of bipartisan support for national workforce policies have made of NSF an institutional solution to complex constitutional and political problems surrounding science and engineering human resources: how to redirect (align) the workforce in fields that the federal government considers important for national survival without interfering with state and local authority over education, while safeguarding the freedom of individuals to choose their professions.[8]

## METHODOLOGY

One can find images of the nation in many locations of U.S. cultural life. For many important reasons, I have selected the popular media as one of the most appropriate repositories of images of the nation. The popular media is the place where threats to the nation are first reported by journalists, politicians, and experts, and usually the only outlet in which both the public and the government express their immediate opinions on the state of the nation. I also use scientific media, such as *Science* magazine, as a location where scientists and science policy makers meet to talk about national problems and their solutions through science and technology. As spaces where different languages meet to talk about the nation, both popular and scientific media provide an appropriate window into images of the nation.

I situate the dominant images of nation around important national events that have had significant scientific and technological dimensions: the launching of Sputnik (1957) which took America into the Age of Science during the 1960s; the Apollo moon-landing (1969) and the emergence of anti-science/technology movements that marked the beginning of a decade of science to meet domestic needs during the 1970s; the rise of Japan as a challenge in the technological marketplace in the 1980s; the appearance in the 1990s of new economic competitors and the emergence of local and global markets that drive the need for a flexible workforce, and, most recently, the terrorist attacks of September 11, 2001. From mainstream popular and science newspapers and magazines, I have collected media representations of the nation, paying particular attention to calls for science and/or technology, and hence for specific types of scientists and/or engineers, to save the American nation from a perceived threat. For the most part, the resources used for this book are accounts from top-level scientists, politicians, and academics and their beliefs about what science and technology can and cannot do to help resolve national needs.

I follow the dominant images of the American nation into official statements made by different institutions and actors who try to define problems and their possible solutions in their own terms. For these, I have used statements found in such documents as oral and written testimonies during congressional hearings, blueprint reports suggesting national actions, and popular and academic articles. All of these have been composed by groups trying to define a problem, to propose solutions in their own terms, and to seek government and public support. Given the NSF's unique mandate to supply the nation with scientists and engineers, I pay particular attention to its statements regarding the solution of national problems. I have collected and analyzed more than 200 reports from the last 50 years from groups and organizations representing government, academia, industry, the professions, and coalitions among these. These documents constitute the archive that I describe and analyze in order to determine the emergence of a discourse.

As Bruce Bimber and David Guston have argued, "the legislative processes, which unfortunately receive little attention in studies of science policy, reveal directly competing conceptions of how science should operate and of what utility science affords society," and add that "the U.S. Congress is a good place to examine how important political institutions are in the shaping of science" (Bimber and Guston 1995, p. 559). Following their insight, I locate the construction of official knowledge around legislative action that follows significant national events in which different groups and actors struggle to define the problems and to propose their solutions as related to scientists and engineers. Inside legislative forums, usually a series of congressional hearings that sometimes conclude in the passing of legislation, I analyze the process of knowledge construction as a power struggle among competing groups trying to gain the control of the problem and its solutions, how this process takes place within the limits of discourse, and how actors are challenged by the dominant image of nation. Nowhere is this struggle better represented than in congressional hearings dealing with NSF's role in solving the problem, particularly those hearings considering proposals for educating scientists and engineers to address new national needs. Therefore, I have collected and analyzed a large number of transcripts of congressional hearings and reports which specifically address the role of NSF in solving national problems by helping educate and train scientists and engineers.

I also conducted and analyzed twelve personal interviews, mostly of government officials, in order to excavate the struggles, past and present, during the construction, legitimation, and transformation of one of the most recent official knowledges and technologies of government institutionalized at NSF: the science and engineering pipeline. These interviews reveal some conceptual limitations of constructivism for they reveal a number of subjugated knowledges or silent voices that traditional constructivist accounts have ignored such as the role of university staff in reproducing the pipeline. The interviews illuminate the different interpretations of official knowledge and of government technologies, thereby showing that closure is not final as is the case with the science and engineering pipeline.[9]

Finally, I follow the trajectory of official knowledge, from its construction to its materialization in programs and policies in education and human resources at NSF, paying particular attention to changes in its budgets, programs, and specific target populations.

## DESCRIPTION OF CHAPTERS

Chapter 2 describes and analyzes how, in the 1960s, the image of the American nation under threat by Soviet science, combined with a discourse of *elite science,* shaped the policymaking process that positioned NSF as the institutional mechanism to make the scientists required for the "Age of Science." At the end of the 1960s, the Apollo moon-landing symbolized the triumph of U.S. technoscience over the Soviet Union. Meanwhile, the Vietnam War, the energy crisis, the awareness of environmental degradation and social and racial inequalities defined the face of a new threat to the American nation. This resulted in an image of the nation facing domestic social and environmental problems and a new discourse of *appropriate science and technology* for society. Chapter 3 describes and analyzes the 1970s, a decade of policymaking to educate scientists and engineers to solve domestic social and environmental problems. In the 1980s, with Japan as a new economic threat and the Soviet Union as a continuing military threat, an image of the American nation emerged that featured a discourse of *technology for economic competitiveness*. Chapter 4 explores how, in the 1980s, powerful actors, many inside NSF, called for large numbers of scientists and engineers for both economic competitiveness and national security to fight this double threat. The chapter examines how some of these actors, in order to count and predict the numbers of scientists and engineers required by the new challenge, constructed a model: the science and engineering pipeline. Chapter 5 traces the emergence of a new image of the American nation challenged by global competition in a multi-lateral world and a discourse of *flexible technoscience* that entered into NSF and its programs. NSF's education and human resources policies and programs shifted from producing large numbers of scientists and engineers "to beat the Japanese" towards making scientists and engineers flexible enough to rapidly adjust to changes in knowledge production, dissemination, and application in new global scenarios. The conclusion chapter summarizes the main arguments and findings of the book, provides an update of the current status of science and engineering education policy, suggests further areas of research, and addresses the question of policy learning, i.e., with each successive policy initiative for scientists and engineers' human resources, was anything learned from the last such situation? Do policymakers mainly respond to images of the nation under threat, rather than opportunities? If so, how does this affect the continuity of what is learned?

## NOTES

1. See for example, Graham Burchel's analysis of liberal government and techniques of the self (Burchell 1993), Barbara Cruikshank's study on self-government and self-esteem (Cruikshank 1993), Peter Miller and Ted O'Leary's study of accounting and the construction of the governable person (Miller and O'Leary 1987), and Nikolas Rose's analysis of the shaping of the private self (Rose 1990).

2. A criticism of this "political anemia" of social constructivism was first made by Langdon Winner (Winner 1991).

3. See for example, "Crisis in Education"(1958), "Educators Upset by Soviet Stroke" (1957a), "Satellite Called Spur to Education" (1957b).

4. See Cooper 1989; Holden 1989; Milbank 1990; Engardio 1994; Greenberg 1995; Magner 1996.

5. There are two other major federal agencies with appropriations for education larger than those of NSF: the U.S. Department of Education, whose programs are not specifically targeted to science and engineering, and the National Institutes of Health, whose education programs are targeted specifically to the biomedical sciences.

6. Throughout the book, I have used the language of the times to show, as accurate as possible, what was on the minds of actors and groups engaged in policymaking. Hence, by using the word "manpower" I do not intend to dismiss the valuable contributions of women in science and engineering, just to show its wide usage in policymaking until the 1980s when the language changed to "human resources" or "workforce."

7. These are the titles of the programs in the 1960s. Today we find most of the education programs under the Directorate for Education and Human Resources (EHR) and the manpower programs under the Division of Scientific Resource Studies (SRS).

8. Other federal agencies, such as NASA and NIH, also make claims about the promises of future returns if the government invests in them. For example, this was the case when NASA tried to justify the Viking-Mars program or the reusable space shuttle program in the early 1970s. Also, both NASA and NIH had implemented educational programs, particularly graduate fellowships, to create specialized scientists and engineers for their space and biomedical programs. However, these agencies do not have the mechanisms to locate, identify, manage, and project larger segments of the population of free citizens into different fields of national interest to the extent that NSF does. Since the 1950s, NSF has had such mechanisms, referred to earlier as technologies of government, in the form of the National Register and Manpower Studies Program, now called Scientists and Engineers Statistical Data System (SESTAT). In the last 10 years, NSF has coupled these population management mechanisms with its education programs, aiming at producing the future numbers and quality of scientists and engineers that national needs require.

9. I am aware of the limitations that the lack of interviews with scientists and engineers bring to my account. Including such accounts could provide a more comprehensive description of how the cultural processes of policymaking analyzed here have actually influenced the individual lives and careers of scientists and engineers. I intend to include these kinds of accounts in future developments of this work.

## *Chapter Two*

# Sputnik Threatens America: Making Scientists for the Cold War

The launching of Sputnik brought significant changes in the image of the American nation, mainly by shifting an anti-communist sentiment from the political into the scientific and educational arenas. McCarthyism had become a thing of the past, but only to be redefined as a fight against communism by science and scientists. A group of scientific elites seized upon America's fears and called for science to save the nation. Government action immediately followed, dramatically increasing financial support for science and science education. Consumerism, the main feature of the "American Way of Life" in the 1950s, now took second place to science, which the government began promoting as a "way of life" (U.S. Department of Health 1961). This chapter analyzes how an image of the American nation now under threat by Soviet science challenged policymakers who responded by positioning the NSF as the institutional mechanism to make the scientists required for the "Age of Science" in the 1960s.

## THE EMERGING OF A NEW IMAGE: COMMUNISM THREATENS AMERICA

During the Cold War's early years, an image of the American emerged, challenging the government and the public to defend the U.S. from communism, and producers and consumers to produce and consume a large output of goods and services generated during the bountiful post-war economy. In response to this challenge, the government invited the American public to launch a war against communism. As Whitfield tells us, "[v]igilance against communism was a national priority during the darkest days of the Cold War, from the late

1940s until the mid 1950s. Abroad, the government mobilized alliances and vast military resources to combat Soviet expansionism . . . [at home, popular media] consistently hammered the theme of an enemy within, working to subvert the American Way of Life" (Whitfield 1991, p. vii). Corporate America and mainstream consumers entered into a holy relationship of mutual production and consumption. Manufacturers produced endless quantities of labor-saving devices and sold them to buyers who sought instant status and gratification. Corporate America promoted images of individual consumerism as symbols of good citizenship and family virtues. Arthur Bec Var, head of appliance design at General Electric in the U.S. put it clearly, " [in the 1950s] . . . more emphasis is placed on home and family living . . . larger families and lack of servants have necessitated as many automatic helpers in the home as possible. By this investment in mechanical servants, an individual can show how he is providing for his family" (quoted in Forty 1986, p. 220). Consumerism came to define the "American Way of Life." Whitfield adds, "Credit cards were launched with the first Diners Club card in 1950, repudiating the ancient commandment of frugality and encouraging immediate gratification instead . . . [t]he bounties pouring forth from American factories and laboratories, made available in such profusion in stores and markets, had become perhaps the chief ideological prop—the most palpable vindication—of 'the American Way of Life'" (Whitfield 1991, p. 71).[1]

American higher education welcomed this challenge of a nation free to produce and consume and willing to fight communism. After WW II, the federal government implemented educational programs that assisted mainstream, middle-class individuals to realize the "American Way of Life" while instilling civic virtues. In return, the American individual ensured his own well-being, along with that of his family and nation. The most ambitious of these initiatives was the G.I. Bill, which became a legislative mechanism to bring the mainstream in line with the needs of the nation. Through the G.I. Bill, the federal government gave war veterans the opportunity to integrate back into civilian society and realize their capabilities as productive citizens. Supported by the G.I. Bill, approximately 7,800,000 veterans went to school (Emens 1965). Most of them studied liberal arts, humanities, social sciences, and business. This legislation brought renewed attention to progressive education in particular, with its emphasis on vocational education (homemaking, health, commercial English, auto mechanics) at the expense of other intellectual endeavors such as basic sciences and foreign languages (Lora 1982).[2]

With the Chinese revolution and the outbreak of the Korean War, Cold War tensions increased in the early 1950s. The focus of education began to move away from personal and community goals towards national goals. For the first time, the pervasive image of America under the threat of international communism seeped into American classrooms. Some educators began viewing educational institutions as instruments of national security (1951). Hence, schools needed to educate in "moral and spiritual values, teach democracy, increase devotion to public welfare [and] make 'efficient producers' [as] a 'resource' to be used by the government for national purposes." Likewise, institutions of higher education had "to lift 'each student to his highest capacity to contribute to the nation's strength'. . . They would contribute to the national defense by educating a major portion of the officer corps and by producing a young population that was unflinchingly anticommunist" (Lora 1982, p. 237). During the years immediately preceding Sputnik, some powerful actors brought the challenges of the America under threat to the forefront of educational policy making. A scientific elite, including Vannevar Bush, James Bryan Conant, and James Killian, stepped forward to oppose the instrumentalism and collectivism of progressive education while favoring the development of a [national spirit] through the "deliberate cultivation of the ability to think" nurtured by basic disciplinary education, particularly in the basic sciences.[3] I call this group of selected individuals, and the ideas and values they promoted, *scientific academism*.

## Scientific Academism and Education

In his account of the role of science in American education, Scott L. Montgomery has labeled the emergence of this elite after Sputnik as "the return of academism." He defines academism as "the oldest educational philosophy in modern Western culture . . . [which] grew out of a system of higher learning originally intended to strengthen and perpetuate monarchical government and its elite servants, by training a stratum of 'higher individuals'" (Montgomery 1994, p. 21). I have added the word "scientific" to Montgomery's definition of academism because this group of extremely influential scientists proposed to strengthen, perpetuate, and save the American nation and its government by educating an elite group of citizens in the basic sciences. The newly educated elite would come to embody the very concept that the scientific academists themselves represented: scientific experts having significant power in both academia and government. Their locations in both government and academia differentiated them from academic scientists who were strictly located inside academia. For example, Vannevar Bush was President of the MIT Corporation and became Director of the Office of Scientific Research (OSRD) and Science Advisor to President Roosevelt. James Conant, the "high priest of Academist policy," was President of Harvard University before becoming NSB's first chairman in 1951. James Killian was President of MIT before becoming Science Advisor to President Eisenhower after Sputnik. These scientific academists worked to ensure the nation's survival through a newly educated elite that would, during times of national need, move from the academy into high-level positions in the government. If successful, the scientific academists would connect the nation to the academy, not only through their own presences and their networks but by promoting and popularizing the idea of elite education for national security. Scientific academism became the most influential group in the construction of official knowledge for postwar science policymaking, particularly for the creation and development of the NSF, its science education and its manpower programs.[4]

Occupying the top positions around and within the NSF, scientific academists began reinterpreting the NSF's mandate on national security.[5] This group of elite scientists used the NSF's seal of approval to publicize the dangerous effects that progressive education was having on national security. Influenced by Conant, President of Harvard and Chairman of the NSB, the NSF commissioned Nicholas DeWitt, of the Russian Research Center at Harvard University, to conduct a comparative study of U.S. and Soviet scientific and engineering manpower (De Witt 1955). With the full endorsement of the National Academy of Sciences and the NSF's director Alan Waterman, the report criticized the products of progressive education, claiming that the U.S. was producing too many businessmen, lawyers, and humanities scholars and not enough scientists and engineers. The report claimed that even though the U.S. had larger enrollments in higher education than the Soviet Union, particularly in the areas associated with progressive education, it fell behind in the production of scientists and engineers:

> Graduating classes in the Soviet Union are today [1954] still substantially smaller than those in the United States. . . . There is , however, a radical difference in the composition of these graduating classes in the two countries. This difference in the graduating classes represents a reflection of the emphasis placed upon the training in specific fields in the two countries . . . [i]n the Soviet Union over 60 percent of the graduating classes were composed of engineering and other natural and physical science majors. . . . In the U.S., less than 25 percent of the graduating classes were in the[se] fields. . . . Therefore, . . . the Soviet Union, with substantially smaller total graduating classes, produced more professionals in the various technical and scientific fields than the U.S. . . . At the same time, of course, the Soviet Union had a drastically small number of "other field" graduates—in the humanities, the social sciences, and the liberal arts—which represents 65 to 70 percent of all U.S. graduates . . . (De Witt 1955, pp. 167–9).

Scientific academism now had in its hands an important official statement with a clear message: the Soviet Union produced more scientists and engineers than the U.S. But even with the powerful endorsement of scientific academism, De-witt's 375-page report, published in 1955, received little government attention outside science policy circles previous to the launching of Sputnik. Although it gained access into the policy arena, the report was not yet able to influence policymaking and action. As U.S. House Representative Albert Thomas (D-Tex) said during the NSF appropriation hearings in 1956:

> This little book, *Soviet Professional Manpower*, I have read word for word . . . and after reading it completely reversed my thinking. . . . Of course we do not have to tie [with] the Russians by any means, but we found out what Russia is doing. . . . If this [book] is true . . . in another five or six years they are going to be ahead of us. Lord help us if they ever reach the point where they are ahead of us, and they are too close to us now(quoted in Krieghbaum and Rawson 1969, p. 187).

Up to this time, the government's attention on manpower had been focused on developing scientists and engineers for atomic energy. Confident with a number of "firsts" in the use of atomic energy, but worried that the Soviets were catching up, the Joint Committee on Atomic Energy of the U.S. Congress held congressional hearings on manpower throughout 1956 to assess how the federal government would initiate "a crash program to increase swiftly and steadily the number of adequately trained American scientists and engineers"(U.S. Congress Joint Committee on Atomic Energy 1956, p. v). U.S. House Representative Melvin Price (D-Ill), Chairman of the Subcommittee on Research and Development of the Joint Committee, understood the potential danger as defined by DeWitt but realized that Americans had not yet come to see this reality:

> It should be no secret that the U.S. is in desperate danger of falling behind the Soviet world in a critical field of competition—the life and death field of competition in the education and training of adequate numbers of scientists, engineers, and technicians. But although it is not a secret, the facts have not sunk into the public mind (Ibid., p. v).

In pre-Sputnik America, a report like DeWitt's, even if endorsed by scientific academism and acknowledged by Congress, could not influence policy until a new image of the American nation emerged and required science education for national survival. For now, the NSF was a small agency, not even considered important in developing manpower for atomic energy. Even after the Joint Committee recognized that the emphasis on creating of this type of manpower "must be on federal leadership because nothing else will do," the NSF was not considered a plausible institution for solving this task. Eventually, with two Soviet satellites up in the sky and an image of the American nation calling for science education, DeWitt's report, re-published in 1961 and now 836 pages long, became a significant piece of the official knowledge that guided post-Sputnik manpower policy, and positioned the NSF's manpower and its educational programs at the center of federal efforts to make scientists for the Cold War.

For now, congressional leaders, the President and high-ranked military officials, all influenced by scientific academism, began to describe a nation in need of education to fight communism. They began to see education as the institution that could produce enough scientists and engineers to preserve national security. However, prior to Sputnik, they had not yet made a commitment to *basic science education*. For example, in May 1957, the Joint Committee on Atomic Energy declared that:

> The war which international communism is waging against us has been referred as the "war of the classrooms." It is a war that at present is not being fought with spectacular weapons as guided missiles, but which is employing a proper instrument of civilization seized upon and converted to use as a weapon; namely education (U.S. Congress Joint Committee on Atomic Energy 1957), p. iii).

Such key national figures as President Eisenhower delivered official statements in settings attended primarily by progressive educators and conceptualized education not in terms of individual fulfillment, but in terms of national survival against communism. Addressing the powerful and pro-progressive National Education Association (NEA) in April of 1957, President Eisenhower stated:

> Our schools are strong points in our national defense. Our schools are more important than our Nike batteries, more necessary than our radar warning nets, and more powerful even than the energy of the atom (quoted in Ibid., p. 104).

Statements by scientific academism that spoke specifically of manpower numbers, such as DeWitt's, resonated with the existing military logic that called for large numbers of personnel to be enlisted and ready for combat. The Armed Forces began defining the problem of the Cold War as that of numbers of scientists and engineers. Early in 1957, Lt. Gen Emmett O'Donell, Chief of the Air Force, declared that the following:

> We are losing a war. We are losing it because we are losing the race to produce more and better engineers and scientists than the Communists are doing (U.S. Congress Joint Committee on Atomic Energy 1957, p. 4).

Scientific academism, however, was careful to position basic knowledge in its proper place. The Cold War, as they saw it, was not merely about classrooms and numbers. It was about the best and brightest students being educated for basic scientific research. In the words of NSB's second Director, Detlev Bronk, in his foreword to the NSF's 1956 Annual Report to the President and the Nation:

> New discoveries of natural knowledge and its application at an unprecedented rate have strengthened our military defenses, stimulated our national economy, and bettered human welfare. . . . In this scientific age it is more true than ever before that trained minds are our greatest sources of power and our most powerful weapons. The mission of the NSF is to prepare the minds of young scientists for the high purpose of research. . . . Appropriations for such a purpose are secure investments in the future of our country and in the welfare of its people(National Science Foundation 1957, p. vii).

On the eve of Sputnik, scientific academists were already seeking national consensus on the importance of science education for national security. And "[when] Sputnik finally pierced the thick armor of American pretensions to superiority," Lora argues, the powerful proponents of scientific academism "looked very much like prophets whose time ha[d] come" (Lora 1982, p. 243). However, before October 1957 there had been no national consensus about what kind of education and what level of federal involvement was best for the nation. The American public was still content with liberal education for the mainstream, while the military was trying to define the problem of education as a matter of numbers. Knowing that the Soviets placed most of their professional workforce into engineering and the sciences, Congress, while cognizant of the need to increase numbers in science and engineering, continued to worry about a federal mandate overemphasizing education in specific fields. The Democrat-led Joint Committee on Atomic Energy, summarizing the main legislative question in its agenda during hearings about developing scientific and engineering manpower, asked: "how to protect the Nation against communism while giving the whole population educational advantage in many fields of knowledge not required by the communist dictatorship" (U.S. Congress Joint Committee on Atomic Energy 1957, p. 29). On the other hand, Congress and the President wrestled over the dangers of federal involvement in education. Fearful of federal control of education, H. Rep. Albert Thomas (D-Tex) stated during the NSF's appropriation hearings in 1956:

> The Congress is now wrestling with the aid to education bill. Congress is pretty well divided and the country is pretty well divided on whether the federal government should step into the field of education or whether the field should be left to the States . . . Is the NSF getting into that field of education that has

heretofore been left to the States? Is the Federal government through the Foundation moving into that field? (quoted Krieghbaum and Rawson 1969, p. 189).

Sputnik made the question of federal involvement in education less important. Congressional leaders from both sides of the aisle and the President reached consensus on the importance of science education for the nation's survival. But prior to October 1957, there was no symbolic evidence of Soviet superiority, only numbers. Reflecting on his directorship of the NSF (1950–1963), Alan Waterman recalled where government officials and policymakers turned, in response to Sputnik, for knowledge about initiating science education programs: "when all the soul-searching began after Sputnik, a lot of people remembered DeWitt's book and the emphasis the Russians had been putting on training scientific manpower. . ." (Interviewed in Krieghbaum and Rawson 1969, p. 226). Sputnik would turn DeWitt's book into official knowledge, and scientific academists into "prophets," not to mention the highest-ranked officials to advise the government on scientific matters including science education and manpower. But first, they had to appropriate the image of America under the threat of communism and define national problems and solutions in terms of science and scientists.

## Sputniks, Science Education, and National Defense

Americans' love affair with space technology began at least two years before Sputnik. As early as 1955, advertising campaigns, popular writing, amusement-park attractions, TV shows and volunteer organizations, such as the Moonwatchers, contributed to disseminating space technology among the public (see Parks 1995). However, when Americans witnessed the great success of space technology with the launching of the Sputniks in 1957, they stood in fear of this new aspect of the Soviet threat. The launches brought the image of communism home in a way that not even Sen. McCarthy's paranoia could. With Sputnik flying over American skies every night, its radio signal beeping on TVs and radios deep inside the American household, the threat of communism expanded to new dimensions in the popular imagination. Popular media depicted Sputnik as "the feat that shook the earth" (1957b) and as the Soviet satellite that sent the U.S. "into a tizzy"(1957a).

Sputnik changed the image of the American nation in many ways and different sectors of American society dealt differently with the challenges of the new image. For example, some sectors of industry and the public responded by incorporating the new objects of fear into consumer goods. Restaurants began serving "Sputnikburgers,"—bars sold "Sputnik cocktails'" and Americans began watching TV programs such as *The Jetsons*, *My Favorite Martian* that positioned space and scientific themes right at the heart of the popular imagination. As Lisa Parks tells us, "mainstream American culture commodified and domesticated Sputnik, positioning it within the discourse of American nationalism rather than leaving it to circle the earth on its own accord" (Parks 1995, p. 16).[6]

Scientific academism also responded to the new challenges by selling science to the general public as Sputnik became visible and audible evidence of Soviet communism's proximity to the U.S., its possibility for world expansion, and its superiority in science:

> For the brotherhood of man, this tremendous scientific achievement should have been an occasion of universal pride and triumph, a time of rejoicing. But the grim, sad fact was something entirely different. Because this achievement had been reached, in a torn world, by the controlled scientists of a despotic state—a state which had already given the word "satellite" the implications of ruthless servitude. Could the crushers of Hungary be trusted with this new kind of satellite, whose implications no man could measure? (1957h, p. 37).

Through science, the USSR had solved the technical problems that kept the U.S. under constant threat and surveillance. As one weekly magazine put it: "Russia had solved important problems of guidance necessary to aim its missiles at U.S. targets . . . [and could] watch the U.S. unhindered and [with] deadly accuracy" (Ibid, p. 41).

Scientific academism disseminated a view of science through the popular media as the reason for the Soviet success and also as the only alternative by which the U.S. could fight the latest communist threat. First, the Soviets immediately reaffirmed their success by releasing international press statements letting the world know of Sputnik's significance for the Soviet Union.[7] A.P. Aleksandrov, president of the USSR Academy of Sciences, wrote to the international press:

> The initial decisive step of a new [scientific-technological revolution] was taken in the first country of socialism. In this region of technology, socialism has surpassed capitalism. The scientific-technological superiority of the new, more progressive social order is clear. And if we still haven't passed the most progressive capitalist countries in all regions of science and technology, then in any case their superiority is a thing of the past (Quoted in Josephson 1990, p. 177).

Aleksarov concluded his article by observing that the USSR had more engineering students than the U.S. and by suggesting that training in physics and mathematics was especially important to ensure the USSR's leadership. Without knowing it, Aleksarov was helping scientific academism in the U.S. make their case.

At home, mainstream popular media popularized the views of both Soviet and American scientists. The same magazines that ran countless ads for labor-saving products in the American home began writing editorials that questioned these products' usefulness in the race to world supremacy:

> The central fact that must be faced up to is this: As a scientific and engineering power, the Soviet Union has shown its mastery. The U.S. may have more cars and washing machines and toasters, but in terms of the stuff with which wars are won and ideologies imposed, the nation must now begin to view Russia as a power with a proven, frightening potential (1957e, p. 41).

In a field-by-field survey of all Russian sciences, scientific experts from all over the world agreed that although the USSR lagged in some fields of applied science, it led the U.S. in most fields of basic scientific research. "What is important about the Russian satellites is the base of science beneath them," the survey article asserted as it went on to ask world experts, "but where do the Russians stand over-all, in the many studies lumped together as science?" (1957a, p. 109) One expert pointed out, referring to mathematics, "Today they are probably ahead of the West in this science [but] the Russians may lag in the applied mathematics of automation and computers." Another expert on solid-state physics asserted that "their basic research is as good as any to be found in the West, and this somewhat counterbalances the fact that Russia has fallen two to five years behind in application" (Ibid). And through the popular media, scientific experts defined the problem in terms of basic science. While lagging in most fields of applied science, the Soviets accomplished Sputnik through basic science, more specifically the mathematical and physical sciences. Now the position of *basic science* in the survival of the nation became very clear. As that weekly magazine concluded:

> The recent Russian advances are not merely isolated technological breakthroughs; they are the result of a long-term emphasis on basic research which is the great strength of Soviet science. An estimated 20,000 Soviet workers annually enter this field, where great discoveries are made (including weapons). By contrast the U.S. increases its basic research staff by fewer than 10,000 workers a year, concentrating instead on applied research leading to the better air-conditioner and the noiseless commode (Ibid, p. 114).

These media depictions of the Soviet threat gave a clear message: basic science had become a weapon to impose political ideologies, "the stuff with which wars are won and ideologies imposed," as *Newsweek* called it. Scientists working on basic scientific research became the new soldiers in the "War of Science." And figuring out "how to mobilize" scientific brainpower became a matter of national survival (1957h). Interestingly, both the popular media and the scientific academists excluded engineering and engineers from their efforts to save the nation. For example, a special issue of *Life* magazine that immediately followed Sputnik showed scientists teaming up with armed forces and business to "map the future [and] to produce missiles" while engineers helped build a dam in Pakistan.

Sputnik helped scientific academists to place not only basic science at center stage but also the federal agencies in charge of funding basic scientific research and education. As MacDougall tells us, "Sputnik became a catalyst for various governmental agencies to sell their particular programs as cures to the presumed ailments of American life that contributed to the loss of the space race" (Quoted in Parks 1995, p. 16). As another group of leading scientists criticized the state of affairs of American science, Dr C.C. Furnas, former Assistant Secretary of Defense for Research and Development, concluded that

> in order to go back and win the race for scientific supremacy, we must revise our naive attitude toward basic research . . . And Congress, now that it has created the National Science Foundation, should have the courage to vote in the funds it needs to carry out many important programs (1957g, pp. 22-3).

Scientific academists defined the new problem facing the American nation not only in terms of basic science and scientists but also in terms of science education. The popular media, after publishing professional educators' early reactions to Sputnik, then turned its focus on comparing Soviet and American educational systems.[8] Although, since 1955, Dewitt's report had contained the most comprehensive USSR-U.S. comparison and had illustrated the U.S. shortfall in numbers, the popular media had better success in shifting the national attention to science education. In early 1958, *Life* magazine ran a 5-part series on the "Crisis in Education" that explored every aspect of "the field of battle for future brain power—the U.S and the Russian schools." The first article of this series, referring to the champions of scientific academism, opened by stating that "for most critics of U.S. education have suffered the curse of Cassandra—always to tell the truth, seldom to be listened to or believed. But now the curse has been lifted. What they were saying is beginning to be believed. The schools are in terrible shape. What has long been an ignored national problem, Sputnik has made a recognized crisis." The article continued criticizing the outcomes of progressive education, claiming that

> in their eagerness to be all things to all children, schools have gone wild with elective courses . . .where there are young minds of great promise, there are rarely the means to advance them. The nation's stupid children get far better care than the bright. The geniuses of the next decades are even now being allowed to slip back into mediocrity. . . . There is no general agreement on what the schools

should teach. A quarter century has been wasted with the squabbling over whether to make a child well adjusted or teach him something (1958, p. 26).

The series of articles concluded with James Conant's blueprint for high school curriculum, the first published statement of his overall idea. In this plan, the article claims, Conant presents "the essence of a significant and workable answer to the basic problem of U.S. education: how to raise its intellectual quality—giving the best minds a chance for unhampered intellectual growth—and yet still preserve its traditional democratic ideas" (Ibid.). This series of articles also provides a good example of how scientific academists responded to an image of the American nation under communist threat by developing a discourse about science and science education as the saviors. Exacerbating the fears of Soviet expansion and domination, they defined the problem as such: a nation in danger, which only basic science and top quality scientists could save. Conant's proposal is the best example of how scientific academists also defined the solution: a meritocratic educational system in which the "bright" students were to be nurtured for scientific careers, the "average" were taught how to build the physical infrastructure, and the "slow" were trained to provide basic services for society. While scientific academists defined the problem, the NSF remained absent for now, only making itself apparent every so often as a possible solution to carry out programs in science education.

## The Image of America Enters Policymaking

The new image of America, with its problems and solutions now defined by scientific academism and popularized by the mainstream popular media, flowed into official statements on educational and manpower policy in science and engineering. Early official statements, such as the one by President Eisenhower only one month after Sputnik, but before the overwhelming popularization of science, shows how prior to 1957 the problem was defined as a matter of numbers in the armed forces. But now addressing the nation about Sputnik for the first time, he said: "this [national security] is for the American people the most critical problem of all . . . we need scientists by the thousands more than we are now presently planning to have" (Eisenhower 1957). Official statements made a couple of months after Sputnik reflect how the new image of America challenged policymakers and how they responded by beginning to set the limits and forms of *what was sayable* about education in the U.S.

In the beginning of 1958, some Congressional hearings on "Science and Education for National Defense" held by the Committee on Labor and Public

Welfare of the U.S. Congress exemplify this process. During these hearings, Senator Lister Hill (D-Ala) not only began setting the limits of discourse, but he did it with the help of scientific academists whom Hill enrolled as witnesses. As Barbara Clowse claims in her historical account of the educational crisis following Sputnik, "his [Lister Hill's] remarks set the tone of the entire proceedings. Both the choice and order of witnesses aimed to establish that education and national defense were now inseparable. HEW officials helped Hill's staff locate witnesses who would testify along those lines" (Clowse 1981, p. 83). Lee DuBridge was among Hill's scientific academist allies.[9] According to Clowse,

> DuBridge was to testify later at the House hearings, and while in Washington, he conferred with committee staff members. He reiterated his hope that federal action could raise intellectual standards in both high schools and colleges. He wanted to see promising students identified and rewarded. DuBridge believed that the federal scholarship idea would help if it were based on merit. He had already offered his opinion of the administration bill to Dr. Killian. He told Killian that the program was justified and attacked serious problems. Like other administrators, he thought that it was a mistake not to offer funds for new buildings (Ibid., p. 84).

Senator Hill further shaped the limits of discourse by enrolling powerful actors such as German rocket scientist Werner Von Braun and U.S. Admiral Hyman G. Rickover. Both endorsed education programs that were oriented towards creating a scientific elite. With an impressive lineup of elite scientists, scientific academists, and high-ranked military officials, it became possible for congressional leaders such as Hill to speak of a nation under threat, not just by Soviet communism, but by its science and technology. Chairman Senator Lister Hill (D-Ala) opened the hearings wondering how "the Soviet Union, which only 40 years ago was a nation of peasants, today is challenging our America, the world greatest industrial power, in the very field where we have claimed supremacy: the application of science and technology (U.S. Senate Committee on Labor and Public Welfare 1958b, p. 2). Lawmakers like Hill could now admit that America's presumed position in the world had been shattered, provoking a national self-questioning unprecedented since times of the Depression. According to Lister Hill:

> [a] severe blow—some would say a disastrous blow—has been struck at American self-confidence and at her prestige in the world. Rarely have Americans questioned one another so intensely about our military position, our scientific stature, or our educational system (Ibid.).

If before Sputnik congressional leaders had any doubts about the value of science education to serve national needs, then after Sputnik lawmakers were able to propose policy for science and science education as means of national defense.[10] Chairman Hill opened the door for supporting basic scientific research and "brainpower" as means for national survival:

> We Americans are united in our determination to meet this challenge. We Americans know that we must give vastly greater support, emphasis and dedication to basic scientific research . . .We Americans know that we must mobilize our Nation's brainpower in the struggle for survival (Ibid.).

High-ranked military officials were then able to criticize, as popular media and scientific academism had already done, the "American Way of Life." Consumerism was undermining our new national duties. No one was more explicit on this issue than Admiral Arleigh Burke, Chief of Naval Operations, who during a congressional hearing that compared U.S. and Soviet science education to determine appropriations for the National Science Foundation, said:

> Our country has grown strong in an environment of personal liberty, human dignity and sound moral and spiritual values. Today, however, in our preoccupation with fringe benefits, higher pay for less hours of work, two cars and a motor boat for every family, we run the grave risk of becoming complacent in our position of world leadership, and of becoming indifferent to the realities of the hard competition we face from the Soviets (U.S. House Committee on Appropriations 1960, p. 4).

Proposals for the federal government's intervention in education were acceptable as long as they would not interfere with the constitutional rights of states and localities. Chairman Lister Hill (D-Ala) recognized the inevitability of federal intervention and the potential for conflict with states and localities:

> The particular task of this committee is to consider how best to stimulate and strengthen science and education for the defense of our country and at the same time preserve the traditional principle, in which we all believe, that primary responsibility and control of education belongs and must remain with the States, local communities, and private institutions (U.S. Senate Committee on Labor and Public Welfare 1958b, p. 2).

What the emerging limits of discourse did not allow was to speak about the right of the federal government to tell its citizens where to work, even if it was on the best interests of the U.S. As Senator Hill put it:

> the Russian under their system tell a scientist or engineer where he is to work and what he is to do. He has to do what they say or get a trip to Siberia. Under

> our free system we do not seek to tell anybody what to do. If they want to work with a tobacco company, of course, they are free to do that. So it makes it all the more important for us, because we do not want to control what people do, that we have a sufficient number [of scientists and engineers] to do these jobs for defense, for teaching of our youth (Ibid., p. 225).

Congress dismissed those bills that proposed centralized actions, such as the proposal to create a U.S. Science Academy to train scientists and engineers in a centralized location, in favor of actions that did not compromise individual liberties or infringe on state or local governments. As Dr. Harry Kelly, Associate Director of Educational and International Affairs at the NSF, put it:

> Education in the Soviet Union is completely designed for purposes of the State . . . and the function of education is to produce the needed servants. The Soviets, in planning the future of their State determine the needs for people in different professions and thereby determine the number of students in each area. In the free world, education is directed towards the development of the individual (U.S. House Committee on Appropriations 1960, p. 14).

Educators dismissed any attempts to apply any part or concept derived from the Soviet educational system. Testifying on behalf of the National Education Association, Mr. William Carr said:

> The Soviet system of education is set up to serve a totalitarian society in which assignment to work is determined by the Government, the allocation of manpower is subject to centralized control. . . . Given our basic economy and our basic values, it is probably as bad a system of education as we could get. I think we ought to go ahead and set up an American system of education, as we have been doing over these years, and not try to worry too much about what the Soviet Union, or any other country, does. If there would be any single rule about the Soviet system of education that we ought to adopt, I would say let us be as different from it as possible (U.S. Senate Committee on Labor and Public Welfare 1958b, p. 477–8).

On September 2, 1958, President Eisenhower signed the NDEA, the National Defense Education Act, was signed into law.[11] It set the limits of what was sayable about the federal government and its role in education to meet the new national needs. According to James Killian in his memoir as First Special Assistant to the President for science and technology, "the bill [NDEA] skillfully avoided the church-state issue and other issues that had earlier proved to be roadblocks to the federal support of education. Not only did it help to strengthen education in both pre-college schools and the colleges by providing new funds; it set the stage for subsequent congressional actions that were to bring the federal government into a whole new relationship to the educational system" (Killian 1977, p. 196). New congressional actions included substantial appropriations for the NSF's science education programs. However, congress itself did not define the NSF's programs. They were defined by scientific academists in both the President's Science Advisory Committee (PSAC) and in top positions at the NSF.[12]

## HIGH QUALITY SCIENTISTS TO SAVE THE NATION

According to Foucault, in addition to setting the limits of what is sayable, discourse also shapes the way in which the relationship between discourse, speakers, and its destined audience is to be institutionalized (Foucault [1968]1991). Scientific academism shaped the way in which *discourse* (science as the savior of a nation under the threat of communism), *speakers* (the scientific academists, and, most recently, the lawmakers), and its *destined audience* (the American public) became institutionalized. Following Sputnik, out of a large number of legislative proposals on education, two institutional solutions emerged: increased budgets for the NSF's science education and manpower programs, and the National Defense Education Act (NDEA) of 1958.

The emergence of the NSF and NDEA as solutions, can be explained by the limits imposed by the discourse. These two solutions not only complied with the limits of *what was sayable* but represented the visions of education that scientific academism had for the U.S. For example, given the reluctance of government to replicate anything similar to a centralized state like the USSR, neither a powerful centralized Department of Education nor the U.S. Academy of Science, envisioned as a centralized national university, became possible.

Under the jurisdiction of the U.S. Office of Health, Education, and Welfare, the NDEA provided funds to several kinds of educational programs: student loans; science, math and foreign language instruction; NDEA general fellowships; training institutes for counselors; language development research; new educational media; and vocational programs. Historians of education claim that the NDEA was the major legislative event in education following Sputnik (Clowse 1981; Lora 1982). Peter Dow goes as far as claiming that the NDEA was "the most sweeping federal education legislation in the nation's history" (Dow 1991). Although the NDEA had major impact on general college education, it had a relatively minor impact on science education and manpower when compared to the NSF. The NDEA was oriented mostly to undergraduate loans and fellowships in all areas, and to increasing education in and providing facilities for foreign languages. For example, during FY 1959–63, only 32% of NDEA funds for undergraduate fellowships went to the physical and natural sciences and 10% went to engineering. Fifty-eight percent went to the social sciences and humanities. Although most of the NDEA's activities were modeled after the NSF's existing programs, the NDEA's only legislative link to the NSF was through its authorization of the Scientific Information Service Program. Hillier Krieghbaum and Hugh Rawson describe the difference between NSF and NDEA programs in terms of executive mandate:

> The recommendations made by President Eisenhower in his 1958 message reflected agreements between officials of NSF and the Office of Education over the respective post-Sputnik roles of the Foundation and the Office in the Federal educational establishment. As described by the President's message to Congress, NSF's domain was to include those programs which "deal exclusively with science education and operate mainly through scientific societies and science departments of colleges and universities." On the other hand, the Office of Education would work primarily through the state and local school systems to strengthen both science education and general education (Krieghbaum and Rawson 1969, p. 228).

The difference between NDEA and NSF programs can also be explained in the way each one defined the problems and solution, in terms of both quality and quantity. The NDEA was to provide federal assistance to the general population, or the "average" and "slow," using Conant's terms, mostly at elementary and secondary levels, hence producing large numbers of educated individuals to serve the nation's needs in manufacturing, infrastructure, and basic services. The NDEA was to be managed by the U.S. Department of Health, Education, and Welfare (HEW), the same office of the federal government that also administered health and welfare programs. In short, it was a comprehensive federal education assistance program for the masses. Meanwhile, NSF programs provided federal assistance to the "best and brightest" with high-quality scientific education in order to produce the scientific elite that would lead the U.S.'s basic scientific research through the Cold War. The NSF, in other words, had become the headquarters of scientific academism. In sum, powerful scientific academists like Killian and Conant envisioned NSF science education programs in terms of producing quality scientists to save the nation from the Soviet threat. According to James Killian

> in education, I urged, we should not engage in an academic numbers race with the Soviets. We must not throw quality out the window in order to handle numbers; our shortage today is one of quality as well as quantity. We should not allow the pressure for scientists and engineers to obscure the need for first-rate

> talent in other fields. . . . James B. Conant expressed a similar concern in a telegram that he sent to the President. . . . These various considerations helped to shape the message on education which the President sent to Congress as well as the bill introduced in the Congress in behalf of the administration. One of the most important parts of the President's message was that the programs of the National Science Foundation for research and for science education should be increased . . . (Killian 1977, p. 193–4).

Other scientific academists in charge of the NSF after Sputnik, such as Alan Waterman (NSF Director) and Detlev Bronk (NSB Chairman) also promoted science education as the solution to create the "best" individual scientists that would save the nation. Originally conceived to "provide opportunities which will carry talented youth, who have already chosen science as a career, to the highest levels of training in engineering and science" (National Science Foundation 1956), NSF science education programs were redefined by scientific academists in the 1960s to "develop capable men and women who [could] be depended upon by the Nation to attain the goals of its scientific endeavors" (National Science Foundation 1960). In the best expression of scientific academism, NSF top administrators single-handedly designed education programs that would reproduce the higher levels in the division of labor as envisioned by Conant. They imagined NSF programs to produce top-level, high-quality scientists who would create the foundations of the new scientific state, and, in addition, enough second-tier scientists and engineers to carry these foundations forward. The NDEA was to produce the rest. As NSF Director Alan Waterman put it:

> our first concern must be to ensure that we are producing in our educational system the *relatively small cadre of top-level*, creative scientists whose genius must provide the foundation and the framework for our total scientific effort. We must then do the best that we can to train *sufficient numbers of highly competent* supporting scientists and technologists to carry forward the work at a sufficiently rapid rate to meet our national needs . . . (U.S. House Committee on Science and Astronautics 1959, p. 113) (Italics mine).

Since this vision depended on a well-defined hierarchy of scientists and engineers, scientific academists decided that it was important to know the location, characteristics, supply, demand, etc. of the population of scientists and engineers (past, present, and future) in order to meet national needs. Having knowledge at its disposal regarding scientists and engineers, the government would then be able to manage these populations without centralized national planning such as that in the Soviet Union. Here, scientific academists further defined the problem in terms of an mathematical equation with an unknown variable that was difficult to find. As Dr. Detlev Bronk, President of NAS/NRC and Chairman of the

NSB put it: "I do think there is a crisis which requires X number of scientists, but the difficulty is no one can say what the numerical value of X is"(U.S. Senate Committee on Labor and Public Welfare 1958b, p. 13). Even if finding the numerical value of X was going to be difficult, now that this value had become relevant to a national crisis, Congress rapidly authorized the NSF to begin searching for X. During congressional hearings on "Scientific Manpower and Education" at the NSF held by the House Committee on Science and Astronautics, U.S. House Representative Edgar Chenoweth (R-Colo) asked NSF Director Alan Waterman: "Do you think this is altogether a matter of numbers?" Resisting the definition of the problem as a matter of numbers instead of quality, Waterman replied: "It is not a matter of numbers, but when you have numbers you can expect a higher percentage, too, of top people, providing their training is equally good" (U.S. House Committee on Science and Astronautics 1959, p. 223). The stage was set for the NSF to begin studying manpower resources in science and engineering, in order to find out how many there were, how many were needed, and how to go about producing that number.

The NSF began to emerge as an institutional solution for the manpower political problem of the 1960s: while respecting the freedom of individuals to choose their own education and employment, how to redirect, educate, and train enough individuals in science and engineering so that a few top-level scientists would emerge to lead the nation's scientific efforts. Now that the relationship between image, discourse, actors, and audience was institutionalized at the NSF, Congress and NSF officials revealed that the more general political problem had two sides: first, finding the number of present and future scientists and engineers, and then providing them with a high-quality education. These two sides of the problem found a home in the NSF's already established Division of Scientific Personnel and Education (SPE).

## Scientific Academists Create Scientists for the Cold War

Immediately after Sputnik, President Eisenhower selected MIT President James Killian as his full-time Special Assistant for Science and Technology and established the President's Science Advisory Committee (PSAC). These appointments opened the White House doors to scientific academism. Robert Kreidler claims that with these appointments "members of the scientific community were given direct access to the President and an established means of expressing themselves on matters of science policy" (Kreidler 1964). PSAC's mandate was to build a national policy for science "in regard to ways by which U.S. science and technology policy [could] be advanced, especially in regard to ways by which they [could] be advanced by the Federal government" (Killian quoted in Kreidler 1964). The PSAC developed policies in three main areas: basic research, facilities, and science education. Their influential role in the last area shaped the forms and the sizes of the NSF's science education programs.

The PSAC's first report on science education, *Education for the Age of Science*, came from its Panel on Science and Engineering Education.[13] This Panel, directed by physicist Lee A. DuBridge, a champion of scientific academism himself, was representative of scientific academism sitting in both academia and in government. The Panel recommended three specific areas in which the federal effort in science education should be placed: 1) recognizing "academically talented" and "unusually gifted" students; 2) developing high-quality teachers; and 3) developing curriculum and courses. Far from being policy, this official statement outlined specific educational objectives for the nation, and one of them stood above all: high-quality science education. As Kreidler argues, "the report did not attempt to make policy. Instead it defined the objectives of policy" (Kreidler 1964, p. 127). If PSAC did not make policy, then who made the policy that guided the largest increment in the science education budget ever seen at the NSF?[14]

The PSAC's recommendations in 1959 found a niche in the existing categories of science education programs at the NSF: Fellowships in Science, Science Teacher Training, and Curriculum Improvement. However, after Sputnik, these programs changed not only in budget size but also in direction. Was it possible that the national crisis brought on by Sputnik was so dramatic that the decisions to direct these efforts were made entirely within the NSF? Was it possible that Congress appropriated $60 million dollars of federal funds to science education without being certain of its direction, hence relying completely on NSF insiders to spend this money as they saw fit? Answers to these questions open windows into the making of policy in the aftermath of Sputnik.

## Quality vs. Quantity within the NSF before Sputnik

Ever since its original conception in *Science—The Endless Frontier*, the NSF Fellowship program has been scientific academism's instrument to ensure high-quality training to a small cadre of top-level scientists. England shows how the various committees that reviewed Vannevar Bush's proposal for fellowships in his 1945 report unanimously agreed on the value of the program to achieve the goals of quality (England 1982, p.13). By 1956, NSF Fellowships were of two kinds: pre-doctoral and postdoctoral. The objective of the pre-doctoral program was "to seek out the most able science students interested in training beyond the baccalaureate degree and to afford them the opportunity to spend full time at the institutions of their choice and in the type of training they desire so that each fellow can develop his potentiality as a scientist to the fullest"(National Science Foundation 1956, p. 72). The postdoctoral fellowships' main goal was "to provide opportunities for scientists who have demonstrated superior accomplishments in a special field to become still more proficient in their respective specialty by studying and doing research in outstanding laboratories" (Ibid.). Scientific academists, such as Alan Waterman and James Conant, in charge of the NSF throughout the 1950s, envisioned and protected the Fellowship program to ensure the making of "higher individuals" in science.[15]

In 1956, President Eisenhower created the National Committee for the Development of Scientists and Engineers. Housed at the NSF and financed from the NSF's budget, the Committee was mostly comprised of representatives from the engineering profession, teachers associations, labor unions, state and local governments, and the social sciences and humanities.[16] The Committee's charter was to "assist the Federal government in identifying the problems associated with the development of more highly qualified scientists and engineers" which included the fellowships (National Science Foundation 1956, p. 17). Throughout its one-year existence, the Committee stayed away from assessing the fellowships and only issued recommendations on how to expand the available supply of qualified technicians. The exclusion of the Committee from assessing programs such as the fellowships implies how jealous and protective the advocates of scientific academism were with their cherished program.

The situation differed for those educational programs, such as the Teachers Institutes, which did not aim at creating "higher individuals" but instead targeted the general population of pre-college students taking science and conceived by scientific academists to manage quality in the teaching of high school science and mathematics. Responding negatively to the appointment of the President's Committee for the Development of Scientists and Engineers, leaders of scientific academism agreed at an NAS symposium that the U.S. should not get into a manpower production race with Russia but should concentrate on raising the quality of scientific education. "They envisioned the Institutes' programs to take care of the problem of high school teaching of science and mathematics" (England 1982, p. 253). However, the Institutes' program was quickly redefined in terms of numbers by its constituency of mostly high school teachers and also by the Office of Defense Mobilization (ODM). "High school science and mathematics instruction must be improved numerically. . . ," wrote the ODM to the NSF, through "courses in science and mathematics which are aimed at high school teachers (and not graduate student) level" (quoted in Krieghbaum and Rawson 1969). Here, the quality of science and math education, as envisioned by scientific academism, was being compromised by large numbers of high school teachers who had not even met graduate-level standards. The Institutes' program rapidly outgrew the fellowships program in budget size but not in prestige. Scientific academism, keen on retaining control of the fellowship program, allowed lower-level regional committees and teachers associations, such as the National Council of Teachers of Mathematics, to define the Institutes' program goals and criteria.

## Quality Prevails at NSF after Sputnik

After Sputnik, scientific academism framed the specific problem of education for the NSF as follows: how to provide high-quality education in the sciences to the "best and brightest" of free-choosing individuals who will save the nation. NSF officials, requesting money for science education by speaking of a threatened nation to be saved by high-quality scientists received no opposition. With scientific academists now in administrative stations at the NSF positioning science education as a matter of national security, their requests for education were taken by the Administration as certified knowledge. Only six days after Sputnik's launch, Alan Waterman, beginning his quest for money for science education, told the National Security Council on Oct 10, 1957:

> This recent event drives home with even more force the conviction that if this country is to compete technologically and maintain military and economic superiority it must maintain its head in science. This is primarily dependent upon the numbers and *especially the quality of our trained scientists and engineers* and the research facilities they need. The country must, therefore, realize the necessity for effective steps toward maintaining progress in basic science and the training of capable scientists and engineers (quoted in Krieghbaum and Rawson 1969, p. 221) (italics mine).

The President's budget and subsequent congressional appropriation for FY 1959 resulted in an increase of 300 percent, to approximately $61 million, for existing NSF educational programs and for the initiation of new ones (National Science Foundation 1958, p. 9). The percentage of the NSF 's total budget ($137 million) devoted to education reached an all-time high of 45% (see figure 2.1).

Never had so much of the NSF's budget gone towards science education. Most of this budget went to funding teachers' institutes and graduate fellowships. With Sputnik in space, scientific academists within high-level policy circles made important decisions regarding program priorities, both target areas and budgets, that shaped the course of scientific research and education. Who were these people and how did they make these decisions? What allowed them to make the decisions in the first place?

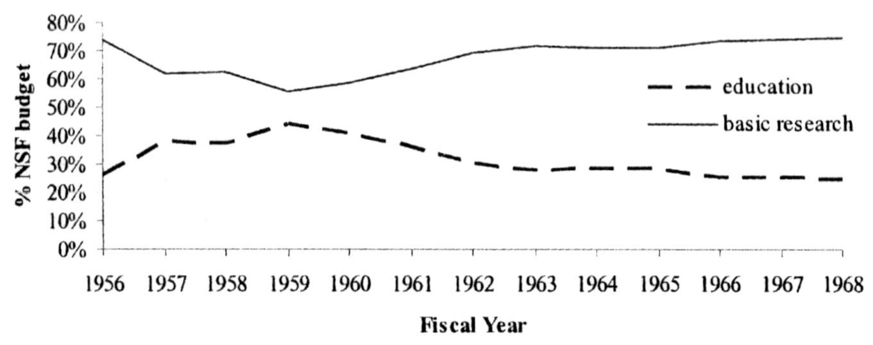


**Figure 2.1. Basic Research and Science Education as Percentages of the NSF's Total Budget: 1960s.**

Source: NSF's Annual Reports from 1956 to 1968.

## Science Teacher Institutes: Providing a Large Pool of Potential Scientists

As we have seen, after Sputnik, scientific academists such as Killian, Conant and Waterman redefined the problem of numbers in terms of quality: the top-scientists who were to lead the nation's scientific enterprise would eventually come from a large pool of potential scientists. With this redefinition of the problem of numbers, the NSF's Division of Scientific Personnel and Education (SPE) became involved in educating large numbers of science teachers with the hope that, from among their students, they would produce a few top-quality scientists.

After receiving direct recommendation from the PSAC to continue and expand the program, the SPE set off to single-handedly redefine the program's guidelines. Howard Foncanon, Special Assistant to the SPE's Division Director Harry Kelly,[17] remembers how the SPE put together the new Institutes program: "We had no real guidelines. The Bureau of the Budget just told us to create a program that would meet the national interest . . . [and] we wrote the new programs" (quoted in Krieghbaum and Rawson 1969, p. 227). Most of the expenditures for this program, which became "the largest single item in the manpower part of the budget" (Ibid.), went to training large numbers of high school teachers with the hope that they would impact large numbers of students from which the nation would draw only the very best. The reach of the Institutes program went even further when it was extended to the elementary level. As Krieghbaum and Rawson claim, "extension of the institutes to the elementary school level was the most significant program development during the 1958–59 period" (Ibid.).

Originally designed to enable the nation's teachers to improve their teaching in order to better stimulate their students to pursue careers in science, the SPE's administrators redefined these programs after Sputnik in terms of national survival, but within the limits of *what was sayable*. Infringement of state and local authority over education still lay beyond these limits. Hence, scientific academists at the SPE redesigned these programs to preserve "the traditional place of state and local governments as managers of their educational systems through the NSF system of support to locally initiated projects [in science teacher training] rather than through the establishment by the Federal government of its own educational operations" (National Science Foundation 1961a, p. 11). Recognizing the accomplishments of the SPE's staff in realizing his visions of scientizing high school education through the Institutes, James Conant wrote in his 1963 book *The Education of American Teachers* :

> *The use of [NSF] summer institutes for bringing teachers up to date in a subject-matter field has been perhaps the single most important improvement in recent years in the training of secondary school teachers* (quoted in Krieghbaum and Rawson 1969, p. 307) (italics in original).

What began with high-level recommendations from PSAC to educate the high school teachers of America in the sciences was really an effort to create a large pool of scientists from which the "best and brightest" would ultimately rise. This was even publicly recognized by the NSF in 1958 when its first annual report after Sputnik stated that "the most serious and urgent problem at the present time in the training of future scientists and engineers is *not to find great numbers of additional students, but to provide a high caliber of training* in science for the competent student who will seek it" (National Science Foundation 1958, p. 49). The NSF further envisioned the making of a select few top scientists through its Fellowships program.

## NSF Fellowships: Making a "Small Cadre of Top-level Scientists"

Consistent with the limits of what was now sayable about the nation, the NSF's top administrators defined the Fellowship program's agenda to strengthen the Nation's scientific potential by "offering aid to graduate students, teachers, and advanced scholars in science, mathematics and engineering according to plans designed to meet the educational needs of individuals." By connecting quality of scientific training to the survival of the nation, they were able to increase their Fellowships budget by more than 100% immediately after Sputnik, from $5.6 million (1527 awards) in FY 1958 to $13 million (3937 awards) in FY 1959 (see figure 2.2). While in 1956 only two kinds of programs existed, by the early 1960s the program developed into seven types of fellowships: graduate fellowships, cooperative graduate fellowships, summer fellowships for graduate teaching assistants, postdoctoral fellowships, senior postdoctoral fellowships, science faculty fellowships, and summer fellowships for secondary school teachers. In 1960, the Fellowship program even included the NATO Fellowships and the Organization for European Cooperation (OEEC) Fellowships to demonstrate to the Soviets the scientific cooperation between the U.S. and Europe.

By 1963, the NSF's Scientific Personnel and Education Studies section produced the most comprehensive projections on the numbers of scientists and engineers for the 1960s: *The Long-Range Demand for Scientific and Technical Personnel* (National Science Foundation 1961b) and *Scientists, Engineers, and Technicians in the 1960s: Requirements and Supply* (National Science Foundation 1963). Lacking its own source of certified knowledge on

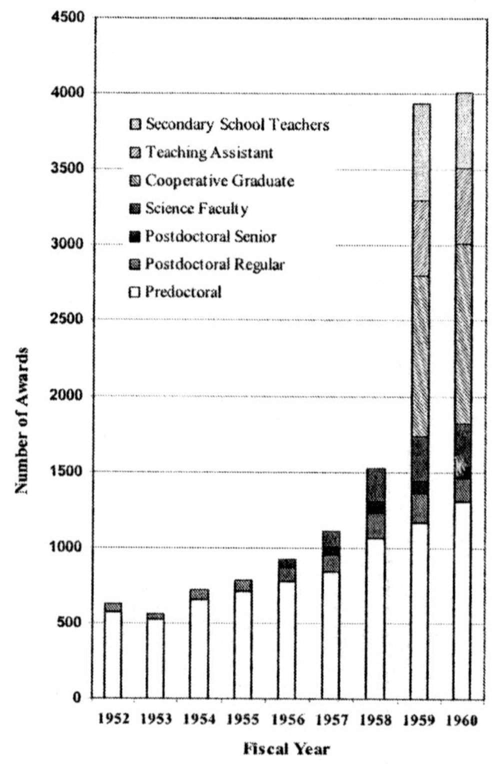


**Figure 2.2. Growth of the NSF Fellowship Program. Number of Awards Offered: 1952–60.**
Source: National Science Foundation 1961a, p. 86.

manpower resources, the PSAC relied heavily on the NSF's official statements mentioned above. Based on its own evaluation of these official statements, the PSAC recommended that the NSF begin a program to support Ph.D.s in engineering. "[O]f all the manpower areas we have evaluated, strenuous and sustained efforts to develop more Ph.D.s in engineering are considered most critical" (President's Science Advisory Committee 1962; President's Science Advisory Committee 1963). This process is indicative of the NSF producing the official knowledge that shapes its own policies. However, for now, scientific academism at the NSF, especially if it was to make engineers, resisted redefining its Fellowship program in terms of numbers in two ways. First, instead of Fellowships in Engineering, the NSF created a numbers-oriented Traineeship program exclusively for engineers, while it kept the Fellowships devoted to the making of only a few quality individuals. Referring to its Fellowship program in its 1964 Annual Report, the NSF states, "from the beginning of its existence the Foundation has stressed the importance of providing support for graduate students and advanced scholars of outstanding ability in the sciences. These individuals represent the backbone of the Nation's scientific potential"(National Science Foundation 1964a, p. 63). A couple of pages later, the NSF justifies the creation of a new program different from the Fellowships

> Recently much stress has been placed on the problem of graduate education in engineering. . . . In 1962 , [PSAC] urged that immediate steps be taken to increase the *number* of master's and doctoral degrees . . . in th[is] discipline. Thus the Foundation is now giving attention to new forms of support of graduate training in certain specialized areas . . . which are known to be in short supply of highly trained manpower. . . . Graduate education for engineers was the first one to receive such supplementary NSF support, since manpower statistics indicated that engineers in particular were in need of advanced training (Ibid., p. 65).

In 1964 the NSF gave out a total of 6,350 graduate awards of which 1,220 Traineeships (20%) went to engineering, 265 Fellowships (4%) also went to engineering, and 1,856 Fellowships (30%) went to the physical sciences.

Second, scientific academism made engineering conform to the demands of science. Criteria for the support of engineering were redefined in terms of criteria of scientific excellence. In 1962, when the NSF established the first Engineering Section and began supporting graduate engineering education, the NSF made clear its criteria for engineering

> The NSF has adopted a policy which clarifies the engineering research supportable by the Foundation by indicating that intellectual pursuits at educational institutions intended to advance significantly the basic engineering capabilities of

> the country are eligible for support by the NSF as basic research in the engineering sciences. *Such work must be of a true scientific nature and not routine engineering practice, and must meet the usual NSF standards of originality and excellence* (National Science Foundation 1962, p. 10) (Italics mine).

## THE MAKING OF A TECHNOLOGY OF GOVERNMENT

Historians of the NSF and science policy have paid very little attention to the NSF's manpower studies. Most of the attention has gone to basic research, where most of the NSF's budget is allocated. So far, this chapter has focused on science education because that is where scientific academism materialized its vision of the kind of scientist needed to save the nation. As we will see, immediately after Sputnik, one of the first requests by both Congress and the President was to find out the number of scientists and engineers available to meet the needs of the nation. However, if scientific academism seemed to be making the kind of policy that the nation needed without such knowledge of the population of scientists and engineers, then why did the federal government request such knowledge?

As Foucault has told us, the ensemble of technologies of government "which has as its target population, as its principle form of knowledge political economy, and as its essential technical means apparatuses of security is what has permitted the [modern] state to survive" (Foucault [1978]1991, p. 100). Post-Sputnik America was no exception. The federal government needed to be certain of its actions, especially if it had to justify continuing them for a period of time. But the government needed more than public acknowledgement of its actions; it needed knowledge of its population to survive. In the U.S. after Sputnik, this knowledge about its population of scientists and engineers, obtained mainly through projections of available manpower and registers of individuals and their characteristics, became essential for national security.

This deployment of technologies of government to project available manpower in science and engineering took place in the early 1960's when the NSF received a request from the President to begin creating knowledge about American scientists and engineers. Sputnik spurred an immediate need for knowledge about available manpower. In 1957, the President through the Bureau of the Budget requested the NSF

> to develop a program for collection of needed supply, demand, employment, and compensation data with respect to scientists and engineers. . . . Because of its functions, its relationship to the National Committee for the Development of Scientists and Engineers, and because it has already arranged for studies of em-

ployment in scientific fields, we believe the Foundation is the most logical agency to undertake such a task (U.S. House Committee on Science and Astronautics 1959, p. 768).

By 1959, the NSF, in cooperation with all federal agencies[18] and scientific organizations[19] engaged in gathering information about scientific manpower, received an additional mandate from Congress to develop a national program of information on scientific and technical personnel. This task was consistent with the on-going re-organization of science and technology activities in the federal government.[20] The NSF became the government's repository for information about U.S. scientific and technological resources, including manpower. As Congress soon recognized, "NSF is the only agency to assemble the available scientific manpower information . . . recording individual scientists by name, profession, and characteristics . . ." (Ibid., p. 335). With an insignificant budget ($1 million, or 0.6% of the NSF 's total),the NSF accomplished this important function of "governmentality" through its Scientific Manpower Program and its two elements, Manpower Studies and the National Register.

## Manpower Studies

As with science education programs and policies after Sputnik, manpower projections made by the NSF in the early 1960s reflected an image of the American nation comprised of free individuals but in need of government action to ensure its survival from the latest Soviet threat. As such, these economic projections, more than just producing numbers about available manpower for current national needs, also divided the national space known as the economy into specific areas of national interest and created population categories according to these areas' new needs.

After Sputnik, the NSF's manpower experts recognized that no projection model existed for this unique situation. Manpower projections for WWII and for the prosperous free market of the 1950s had been made using models appropriate to those national scenarios. In early 1960s, manpower experts told Congress, "no impression should be given that precise future estimates of scientific personnel requirements will eventually be possible [because] the lack of an adequate conceptual framework has been the basic problem" (U.S. House Committee on Science and Astronautics 1960, p. 33). Immediately, Congress authorized the NSF to initiate a study "to develop a systematic methodology for the long-range projection of demand for scientific and technical personnel," mainly "as a national defense resource" (Ibid.). Accordingly, the NSF defined the goal of its Manpower Studies in national terms: "[it] is the central program in the Federal government for the provision of . . . data on the supply, demand, education, and characteristics of the Nation's scientific and technical personnel resources"(National Science Foundation 1960). In the next three years, the NSF 's SPE single-handedly constructed a model that responded to the challenges of the new image of nation (see National Science Foundation 1961b; National Science Foundation 1963). Projections from this model significantly influenced education manpower policy for the remainder of the 1960s.

The new model reflected a shift from perceived national economic needs in the 50s to those required by a nation now under threat by Soviet science. In the 1950s with a free market of goods, services, and labor plentiful after the war, projection models utilized free-market models. Due to salary changes, free-market models took into account adjustments in the supply and demand of labor. In these models, salary differences among occupations played a major role in influencing supply of workers and in determining where they went. The NSF adopted this model in the 1950s as developed by its grantees David Blank and George Stingler who utilized a free-market model comprised of free individuals behaving as economic actors in accordance with economic incentives of a free-labor market. In this model, no notion of centralized or state-directed economy existed (Blank and Stingler 1957). Their model, which superseded wartime models used to project manpower in a government-controlled labor market for wartime needs, now depicted the behavior of free American individuals choosing careers that suited their best interests, usually graduate degrees in science and engineering. Black and Stingler concluded that the market of supply and demand of scientists and engineers had adjusted over time, that there was no labor shortage, and that more students went into graduate degrees, as shown by the increase of the ratio of Ph.D. to B.S degrees. This model depicted the nation's labor sector as made up of two broad areas of national importance, military and civilian industry, and of free-market forces allocating individuals well within these areas: "It appears clear, therefore, that economic incentives have played an important role in attracting science and engineering students toward those fields in which demand has been high and increasing" (Ibid., p. 83).

After Sputnik, however, new national needs required the government to intervene in the supply of manpower. These new needs redefined the two broad areas of the economy into specific areas of national importance. Among these, basic research became one of the most important areas within both civilian and military industries. Within the category of basic research, the subcategories of physical sciences, life sciences, engineering, social sciences, etc. began gaining national attention, and they were ranked according to their contribution to the nation's survival. Subcategories were further sub-divided according to specific disciplines. The government now had to intervene by allocating people into these new areas of national significance; hence it had to act on the supply of manpower. Accordingly, the NSF shifted its projection model to a "fixed coefficient" model, which allowed the government to intervene in projecting the supply and demand of manpower.

Fixed coefficient models rely upon the past trends and the projected future trends regarding the distribution of individuals in education and employment, and the parameters that affect these distributions. The parameters usually selected are those in which the government has a strong influence, such as federal employment and R&D spending. Projected demand is a function of the input to that activity (e.g., R&D spending), its outputs (e.g., scientists required), and assumptions about its productivity trends (e.g., more R&D means more scientists required). Projected supply is based on demographic patterns of the educational and occupational choices of persons entering the labor market. To eliminate the free market's major influence and favor the government's role, projecting supply and demand does not take salary compensation into consideration.[21]

The NSF's new fixed-coefficient model positioned the federal government as the main determinant of projected demand. The model assumed that demand of scientists and engineers was in direct relationship with federal R&D spending and with the government's employment of scientists and engineers. It established a history of the demand of scientists and engineers for known levels of federal R&D spending and for the utilization of scientists and engineers during the 1950s, and extrapolated this into the 1970s. Supply was projected on the basis of population growth, the proportion of college-age population actually going to college, and the proportion of the entire college population going into science and engineering (National Science Foundation 1963). To remedy this demographic determinism, the NSF's model also positioned federal programs, particularly those at the NSF, as solutions to the supply problem by increasing the number of students going into science and engineering.

The new model, which became "the most comprehensive and systematic attempt at quantification of the future of [s/e] personnel . . . in this country," projected a shortage of 170,000 scientists and engineers by 1970 (National Science Foundation 1961b). These projections became the most influential official statements guiding both Congress' and the President's educational and manpower policies throughout the 1960s. For example, Congress conducted its own policy analysis based on this model and these projections, only to find that the NDEA of 1958 had a minimal effect on the post-Sputnik population going to college. Referring to his own study and the findings on the NDEA, Spencer Beresford, Special Counsel to the Committee on Science and Astronautics of the U.S. House of Representatives, asserted that

> all present programs and legislative proposals of this kind appear inadequate in scope and extent. For example, Dr Bolt's study [congressional study using the NSF model] failed to show that the NDEA has had any influence whatsoever on the annual percentage of Americans getting college degrees, that is, its influence has been too small to be seen on the chart. The ideal is economic aid on the scale of the GI Bill after WWII—which is reflected on Dr. Bolt's chart by a sizable 'postwar bulge' (U.S. House Committee on Science and Astronautics 1962, p. 8).

Since no GI Bill came after Sputnik, the NSF, through its science education efforts, remained the only solution to this shortage. We have already seen how the President's Science Advisory Committee also endorsed the NSF projections and recommended specific policies for graduate education in engineering which the NSF reluctantly carried out through a new traineeship program. In short, these projections positioned the NSF not only as the producer of the most legitimate and respected official statements about the future supply and demand of scientists and engineers, but also placed it, through its programs, as the executor of policy to solve the manpower crisis.

The NSF's projections for the 1960s, as legitimate as they were, still had to comply with the *limits of what was sayable*. After comparing DeWitt's study of Soviet manpower with the NSF's projections of U.S. manpower, Beresford, speaking on behalf of the Committee on Science and Astronautics, remained reluctant to compromise individual liberties in order to compete with the USSR:

> Obviously, any comparison of education or of s/t manpower in the U.S. and the USSR must be viewed with reservations. The political, social, and economic institutions—and, above all, the policy objectives—of the two countries are very different. In the U.S., the individual citizen is regarded as an end in himself; in Communist society, he is merely an instrument for promoting the collective good. The United States educates approximately twice as many citizens, at every level, as the Soviet Union. In the Soviet Union, all education is planned and directed by a central authority, which heavily emphasizes engineering, the natural sciences and mathematics . . . (Ibid., p. 4).

Aware that regulating salaries and directing people into critical scientific fields by force were not feasible, the Committee on Science and Astronautics proposed the deployment of government technologies in order to establish control over science and engineering manpower:

> It is probably not feasible, either, to establish economic or physical controls for critical types of scientific and technical manpower. If such controls were adopted, they would probably be voluntary. Short of controls, however, it would help to keep track of the supply and utilization of scientific and technical manpower (Ibid., p. 7).

The NSF exercised such manpower control through its National Register of Scientific and Technical Personnel.

## The Register

The Register's legislative origins go back to the NSF Act of 1950 as "a register of scientific and technical personnel to serve as central clearinghouse for information covering such personnel." Before Sputnik, the Register's categories reflected the wide range of possible categories envisioned as scientific manpower. In 1956, the Register was described as containing "data necessary for an understanding of the adequacy of present and potential supply of science manpower range from relatively abstract psychological problems of creativity to practical questions of numbers now employed in research and development as against estimates of science manpower needs 10 or 15 years from now under a variety of possible conditions" (National Science Foundation 1956, p. 74). Also redefined around the new national needs after Sputnik, the Register's new goal became "to insure that information on the resources of scientific manpower is available, and that individual scientists and engineers with specialized skills can be identified and located as required in the national interest" (National Science Foundation 1959, p. 85). Accordingly, the NSF began to redefine old categories into new ones now considered of national importance. In 1958, one year after Sputnik, the NSF intended "to place the Register program on a more current operating basis and extend the coverage of the Register to new fields of vital importance to the nation, such as rocket and missile technology, communications and electronics, aeronautical science, ceramics and metallurgy" (National Science Foundation 1958, p. 68).

In addition to endorsing scientific academism's vision in which the "best and brightest" students of science would save the nation through basic scientific research, the NSF established the important traits of scientists who were to be collected in the Register. In 1964, the NSF published the first comprehensive study that described U.S. scientific and technical manpower according to the following characteristics: employment status, geographic distribution, level of education, age, years of experience, type of employer, type of work activity, foreign language proficiency, and income (National Science Foundation 1964b). A look at these categories can give us a further sense of who was considered to be a scientist in the 1960s. For example, the average scientist was a 30–year old male from the middle-Atlantic states (usually New York), with a Ph.D., 10 years or less experience in the field, was employed in either academia or in industry, and more likely to know German than Spanish.

This average scientist deserves further analysis because, in many ways, several of its aspects have become ingrained in what we understand as "scientist."

For many years after Sputnik, this was, and still is, the image of "scientist" that we saw in popular literature, movies, TV, oral descriptions, etc. and, for the most part, the person who inhabited science labs and presented papers at professional conferences. However, the NSF took for granted some important traits. I finish here with a look at two of the most important—gender and race. For, although invisible in the 1960s, new images about the nation emerged in the following decades, and scientists' gender and race became contested in different ways. After looking at the concept of "scientist" that emerges from NSF's technologies of government, some questions come to mind: Where were the women and non-whites involved in these scientific efforts to save the nation? Even after congressional committees, the NSF, and policy advisors grew aware of the large numbers of women and racial minorities participating in Soviet scientific manpower, why were American women and racial minorities not included to participate in the efforts? In trying to answer these questions, it is worth looking at the limits of appropriation of discourse that developed around Sputnik.

In determining these limits, Foucault asks us to pose such questions as "what individuals, what groups or classes have access to a particular kind of discourse?. . . How is the struggle for control of discourses conducted between classes, nations, cultural or ethnic collectivities?" (Foucault [1968]1991). As we have seen, powerful actors, most of them scientific academists, with definite ideas of how to save the nation through science defined the problem of science education on their own terms, never allowing other social groups to advance their own images of the nation, their definitions of the problems, or their proposed solutions. Even though DeWitt's analyses of Soviet manpower (De Witt 1955; De Witt 1961) revealed a significant diversity among Soviet scientists and engineers,[22] and Congress acknowledged that "other social groups, such as the Negroes, make a disproportionately small contribution to the national supply of scientific and technical manpower" (U.S. House Committee on Science and Astronautics 1962, p. 5), the NSF projections and Register's categories, all based on limits of discourse set by scientific academism, rendered the categories of gender, race and class invisible. This absence clearly reflects the limits of appropriation established around Sputnik. The American nation was to be saved by a specific concept of "scientist" that the NSF helped create.

## CONCLUSION

Under the symbolic threat of Sputnik, scientific academism successfully appropriated the limits of what was sayable about the nation and its science and education. Located strategically inside both the federal government and elite academic institutions, this group defined a very specific national problem and its solutions, while silencing alternative voices and their proposals and rendering some population categories invisible. The problem was cast in terms of the dominant image of the American nation under threat by Soviet science and science education. The solution: the production of "a small cadre of top-level" scientist who would lead the scientific efforts to save the nation. With pressure coming from other groups, such as the military and the teaching profession, to define the problem also as a matter of numbers, scientific academism successfully resisted number-oriented solutions to a problem they had defined, instead, in terms of quality.

Within the limits of what was sayable and acceptable in the discourse about the American nation in the early 1960s, the NSF emerged as an institutional solution to the political problem surrounding the production of scientific manpower. That is, the NSF became the federal government's solution with which to redirect, educate, and train enough individuals in science and engineering so that a few top-level scientists would emerge to lead the scientific efforts of the nation while respecting the freedom of individuals to choose their education and employment. However, the solution to this political problem required more than money for science education. Given the significance of scientists to national survival, the federal government also needed certified knowledge about the nation's scientific and technical population in order to be certain of its actions regarding science education and manpower. The NSF's manpower studies program became the site from which this certified knowledge was produced. The manpower projection model used by the NSF embodied an image of a nation under threat, in which federal intervention was required but not at the expense of the free-market. The NSF used fixed-coefficient models, which allowed the government to intervene in the supply of required manpower without intervening on salaries. The creation of this knowledge within the NSF would also create an interesting conflict of interest since it was the same knowledge that shaped the NSF's own policies.

Throughout the 1960's, the NSF , by means of its technologies of government —manpower projections and the Register—helped create a stereotype of scientist with specific traits, skills, and specialized knowledge who would save the nation from the Soviet threat. This stereotype continues to influence what we understand by the word "scientist" today.

## NOTES

1. One of the best analysis of consumer culture in the 1950's is Fox and Jackson 1983. For an excellent account of the evolution of consumer-oriented design in the 1950s see Forty 1986.

2. For examples of the influence of Progressive education on American education in the early 1950s see U.S. Office of Education 1951.

3. For the most controversial criticism of progressive education in the 1950s see Bestor 1953.

4. For Conant's elitist ideas on science education, see Conant 1945, 1952. For Rickover's views on academic elitism and national security, and his pleas for more scientists and engineers to lead the war against Soviet communism, see Rickover 1959.

5. For a detailed account of how a small group of scientific academists shaped the policymaking process during the postwar that defined the NSF 's structure and future, see Kleinman 1995.

6. For a feminist analysis of the appropriation of discourse through film and sci-fi literature, see Spigel 1991. For a more general analysis of the appropriation of discourse through business, see MacDougall 1985.

7. For a complete analysis of how Sputnik was viewed in the USSR see Josephson 1990.

8. For examples of immediate reactions from educators to the Sputnik crisis, see New York Times articles 1957b; 1957c; 1957d; 1957f.

9. At the time, DuBridge was president of CalTech. He had been a member of AEC (1946–52), chairman of the Science Advisory Committee of the U.S. Office of Defense Mobilization (1952-56), member of the National Science Board (1950–54 and 1958–64), and member of the National Manpower Council (1951–64). He eventually became Science Advisor to the President in 1969–70.

10. For a list of all bills related to science education at this time, see U.S. Senate Committee on Labor and Public Welfare 1958a.

11. See U.S. Senate Committee on Labor and Public Welfare 1958a; U.S. House Committee on Education and Labor 1961; Clowse 1981.

12. Membership of the PSAC included: Robert F. Bacher, Professor of Physics, Cal Tech; William O. Baker, VP (research), Bell Telephone Lab; John Bardeen, Professor of Electrical Engineering and Physics, University of Illinois; Hans Bethe, Professor of Physics, Cornell University; Detlev W. Bronk, President, The Rockefeller Institute; Britton Chance, Director of Biophysics, University of Pennsylvania; James B. Fisk, President, Bell Telephone Labs; George B. Kistiakowsky, Professor of Chemistry, Harvard University; Edwin H. Land, President, Polaroid Corporation; Emanuel R. Piore, Director of Research, IBM Corp.; Edward M. Purcell, Professor of Physics, Harvard University; Isidor Il Rabi, Professor of Physics, Columbia University; H.P. Robertson, Professor of Physics, Cal Tech; Glenn T. Seaborg, Chancellor, University of California; Cyril S. Smith, Institute for the Study of Metals, The University of Chicago; Paul A. Weiss, Member and Professor, The Rockefeller Institute; Jerome B. Wiesner, Director, Research Laboratory of Electronics, MIT.

13. James Killian, president of MIT and now the President's Science Advisor brought two of his favorite MIT professors to serve in this panel, Jerrold Zacharias, Professor of Physics, and John Buchard, Dean of Humanities and Social Studies. See Killian 1977, p. 196, for his own account of the formation of this elite panel.

14. The NSF's total budget rose from $51 million in FY 1958 to $137 million in FY 1959 while, in the same period, the science education budget went from $20 million to $60 million. This represents an increase for science education from 38% to 45% of the NSF's total budget.

15. For FY 1956, the NSF awarded a total of 925 fellowships, 70% of them in the physical sciences, which included engineering, and 30% in the life sciences.

16. For a comprehensive list of members, see National Science Foundation 1956, p. 19.

17. Harry Kelly was one of the best representations of scientific academism. A colleague of James Killian at MIT, Kelly also worked as Chief of the Scientific and Technology Division of the U.S. Army in the occupation of Japan during 1945–50. Before joining the NSF in 1953 as an appointee of Alan Waterman, Kelly worked for the Office of Naval Research.

18. These included the U.S. Departments of Commerce, of Defense, of Health, Education and Welfare, of Justice, of Labor, U.S. Civil Service Commission, U.S. Selective Service System, U.S. Veterans Administration, and the Atomic Energy Commission.

19. These include National Bureau of Economic Research, the Bureau of Applied Social Research, the NAS/NRC Office of Scientific Personnel, the National Manpower Council, the Engineering Manpower Commission, the Scientific Manpower Commission, and nine major professional scientific societies.

20. This is exemplified with the establishment, in 1959, of the Federal Council for Science and Technology which included the Departments of Agriculture, of Commerce, of Defense, of Health, Education and Welfare, and Of Interior, and from NASA, the NSF , and the AEC. See Executive Order 10807, signed by President Eisenhower in March 13, 1959.

21. For a complete analysis of both free-market and fixed-coefficient models, see National Science Board 1974, p. 5.

22. DeWitt's analyses show significant contributions from Soviet women and national and ethnic minorities to their scientific and technological manpower. For example, women constituted 50% of the science and engineering professionals in the Soviet Union.

*Chapter Three*

# Social and Environmental Problems in America: Making Scientists and Engineers for Domestic Needs

> "To educate scientists who will be at home in society and to educate a society that will be at home with science."[1]

At the end of the 1960s, the Apollo moon-landing symbolized the triumph of U.S. technoscience over that of the Soviet Union. Meanwhile, the Vietnam War, the energy crisis, a new awareness of environmental degradation and social and racial inequalities all defined a new image of the American nation. The image of the American nation shifted from one under the threat of communism to one under the threat of its own domestic problems. If science had just saved America from Sputnik by putting the U.S. ahead in the space race, now science and technology were perceived as threats to America's own social and natural environments. This new image created a tension between a renewed scientific academism that still occupied the NSF's higher ranks and supported basic science, and a bandwagon of progressive groups that wanted to bring the NSF in touch with the nation's social and environmental problems by supporting applied science. Throughout most of the 1970s, as the NSF's responsibilities for funding this new kind of science grew, these two groups battled for budget allocations and programs thereby threatening the NSF's commitment to basic science. Within this context, the NSF's priorities in developing education and manpower shifted from producing top quality scientists for the "Age of Science," to producing scientists capable of solving the new problems that plagued American people, cities and the environment. This chapter looks at a decade of policymaking around NSF to educate scientists and engineers who, in addition to saving the nation from the threat of communism, could solve the problems of society and the environment.

## THE EMERGING OF A NEW IMAGE: AMERICA AT WAR WITH ITSELF

By the end of the 1960s, Apollo 11's landing on the moon symbolized the triumph of technoscience over the Soviet threat in space, and removed the Soviet Union's perceived scientific advantage of over the U.S. It shifted America's perception of the enemy. After this moon landing, popular media claimed that America had regained its supremacy in science and technology over the Soviet Union. As one weekly magazine celebrated one week before the launching:

> Just a few years ago, the U.S. seemed hopelessly outstripped in the space race. It was an uneasy America that watched the Russians score one "first" after another in the skies. Today it is America, not Russia, that stands at the threshold of the greatest space conquest of all—landing men on the moon. For much of mankind, setting foot on the moon has been a dream for centuries. It was left to Americans to fulfill that dream(1969a, p. 1).

But more than just symbolizing the mastery in the realms of science and technology, the moon landing represented a triumph of American capitalism over Soviet socialism. Just two weeks after the Apollo landing, mainstream media summarized the triumph as follows:

> A race won. Here on earth, practical evidence was given that the U.S. has emerged, after a decade of self-doubt, as the most technologically and scientifically advanced of all nations. The U.S. won the race to the moon and threw back into the face of the Soviet Union a boast of seven years ago that Russian space achievements would "demonstrate the great advantage of the socialist system." In the end, it was a society of free men and competitive industry that demonstrated the advantage . . . (1969b, p. 24).

While most Americans celebrated the Apollo success, some questioned not only the usefulness of the event but also the responsibility of the scientific and industrial establishments in social and environmental degradation of the American landscape. For skeptics, putting a man on the moon was a dubious accomplishment in the face of other crises facing America. Mainstream popular media juxtaposed images of both technological enthusiasm and skepticism: "A mixture of awe and uneasiness: That is the reaction to Apollo 11" (1969, p. 71). Reports questioning whether "60 pounds of lunar rock were worth twenty-four billion dollars" accompanied reports that chronicled the medical, environmental and urban crises facing the nation. For example, only one week after the moon landing, *U.S. & World News* published three main stories under the titles "How useful is the moon," "Medical Crisis and How to Meet it," and "The Breakdown of Our Cities." As one editorial summarized: "We embraced the space program and the industrial-technological juggernaut it spawned . . . but the verdict of history may well be that, while the world erupted, we ignored the real challenge and chased a rocket trail to the moon" (1969, p. 71).

Among the stories surrounding Apollo, one became more and more publicized: the effects of a shrinking space industry on American workers. As reports published how "Apollo has absorbed the efforts of a half-million workers, charted the technological course for 20,000 corporations . . . and it has built a dependence on Federal financing into a massive segment of the national economy," others reported the" big let down after a big build up" (1969). Among those workers were, of course, many scientists and engineers.

## Scientific Academism and the Struggle to Redefine Science

In the beginning of the 1970s, the new image of nation brought two responses regarding how technoscience could serve the US. One response celebrated the nation's triumphs in science and technology, as exemplified by winning the space race to the moon, and wanted more of the same. At the highest levels of government and academia, a renewed scientific academism endorsed this celebration and resisted criticisms of the scientific establishment's status quo. In an extensive interview article with the most influential of these scientific academists, one leading national magazine summarized their views on what they called a "senseless war on science" and put in brief their beliefs on the nation and its science as follows:

> Over the last quarter century, despite dips and lags, science and the U.S. economy together have had the longest period of sustained growth, discovery, innovation, and new industry in recent history. While none of this has brought the millennium . . . still it has measurably widened the options and potentials of human life on earth. Perhaps the pinnacle of this period was reached when man stepped on the moon, a feat that will rank in history among the few clear, large, and positive achievements of the last decade, a great human feat that once would have swelled the lyrics of a Homer . . . (1971, p. 89).

The other response was to blame science and technology for the nation's problems. This time the mainstream media, instead of neglecting dissenting voices as it had in the late 1950s, now provided column space for this criticism. In the 1970s, some high-level academics, including scientists, promoted a view of science and technology as capable of destroying the nation. Even some scientists, such as Harvard biologist Everett Mendelsohn, came to criticize science, and raised the possibility that science, as developed after

WWII, had reached both its rational and practical limits. In contrast to media reports following Sputnik, in which scientists elevated science to the level of national savior, reports from the early 1970s depicted not only a polarized nation but also a divided scientific community disputing whether science was savior or oppressor. Reporting the views of science critics, including those of scientists themselves, a special issue of *Time* magazine entitled "Second Thoughts about Man: Reaching Beyond the Rational" reads:

> . . . there has begun to emerge, even within the laboratory, a new fascination with what traditionalists consider the very antithesis of science: the mystical and even irrational. Says Harvard Biologist-Historian Everett I. Mendelsohn: 'Science as we know it has outlived its usefulness' . . . Science did indeed bring forth a Brave New World-of transistors and miniaturized electronics, antibiotics and organ transplants, high-speed computers and jet travel. But progress came at a price. It was the genius of science that also made possible such horrors as the exploding mushroom cloud over Hiroshima, the chemically ruined forests of Indochina, the threat of a shower of ICBMS, a planet increasingly littered with technology's fallout. It is this Faustian side of science, with its insatiable drive to conquer new fields, explore new territory and build bigger machines, regardless of costs or consequences that worries so many critics (1973b, p. 83–4).

Science enthusiasts, including practicing scientists and scientific academists in government, rushed to gain control over the attack on science. They attempted to redefine it in their own terms and through their own media. *Science* magazine, read by science policy makers and academics alike, became a frequent forum through which science enthusiasts, now challenged by a new image of the American nation, shaped a new discourse about science and technology as solutions to the nation's problems. Scientists from all ranks rushed to defend the accomplishments of science and to resist changes to the status quo regarding federal funding of science. As one physicist wrote in an editorial of *Science*:

> How should we defend science against its attackers? To what extent should we change direction so as to work more specifically on the applications of science to the public good? Should we as scientists throw our support behind one or other of the major social and political forces, or indeed behind some other political force? . . . No, the only effective defense of science is through strengthening science itself . . . (Thimann 1970, p. 1).

Just one week later in the same editorial, a scientist trying to defend basic science quoted the voices of the nation's top scientific academists:

> To the extent that society insists that basic scientists do work that is more relevant to present social needs . . . scientists will be less able to work where nature

> appears willing to answer their questions. They may be required to work on relevant questions that perhaps cannot be answered at all at present, or can be answered only with uneconomic use of resources. Thus, excessive efforts to make science more productive in terms of immediate social goals may actually make it far less productive in the long run (quoted in Abelson 1970, p. 1).

On the other hand, science critics, in their own terms and on their own media channels, also attempted to define a discourse about the nation's problems and solutions. Having defined the problem as that of science's unresponsiveness to the nation's needs, some of these critics developed a solution in which the most immediate national problems would be solved by a technology "of a different kind." For example, Professor Robert Heilbroner of the New School in New York, after outlining the nation's most pressing problems—poverty, collapse of urban infrastructure, racism, and environmental degradation—proposed as the main priority for the seventies to be "not for less technology, but for more technology *of a different kind:*"

> The priority then is technological research—research aimed at devising the techniques needed to live in a place that we have just begun to recognize as our Spaceship Earth . . . (Heilbroner 1970, p. 84).

Defining the solution in even more detail, Heilbroner proposed using existing scientists, engineers, and technicians, and training new ones in American research universities in order to carry out this "different kind" of technological research:

> Many people wonder where we can direct the energies of the engineers, draftsmen, scientists, and skilled workmen who are now employed in building weapons systems, once we cut our military budget. I suggest that the design of a technology for our planetary spaceship will provide challenge enough to occupy their attention for a long time. . . . I can suggest aimed specifically at the upper echelons of the educational apparatus . . . that they direct major portion of their efforts toward research into, training for, and advocacy of programs for social change . . . (Ibid.).

Given the emerging image of a nation plagued by social and environmental problems, it became possible for a discourse about science at the service of society to evolve. Even in scientific media, stories of scientists from all ranks putting their service to social problems became possible and acceptable. One editorial, after telling the courageous decision of James Shapiro, one of the most promising molecular scientists in the nation, to leave scientific research for political activism, concludes:

New scientific knowledge will be required to counter the effects of the uses to which current knowledge has been put . . . we wish Dr. Shapiro success in his new career—for today, more than ever before, it is imperative that the fruits of science and technology be turned exclusively to the service of man (Cass 1970, p. 57).

While some scientists proposed science education and research for social change as a plausible solution to national problems, others eager to establish control over this new discourse about science recommended a variety of fixes to the scientific establishment itself, including adjustments to existing governmental institutions (Gershinowitz 1972). One of these moderate reformers, A. Hunter Dupree, one of the nation's most respected historians of government and science, became aware of the potential implications that these attacks and proposals could have with regards to the partnership between government and the scientific establishment. Regarding the role of government, Dupree suggested

on the government side of the partnership, the problem is not to make a paper reorganization , to create a Department of Science, or to disturb the functioning of the science advisory . . . within the White House. It is rather to provide a comprehensive rationale by which the government can continue to support free science in the universities and whatever else it can find an institutional home. . . . Any new arrangement must improve the position of the NSF in . . . emphasiz[ing] the chain of connections, and not the disconnections, between long range basic research and applied science generally, in the interests of both national security and of alleviation of social and medical problems besetting mankind (Dupree 1970, p. 57).

Important science leaders came to accept the new limits established by the discourse about science usefulness to society. As Bentley Glass, President of the AAAS (1969) and Vice President of SUNY at Stony Brook, said: "Science PhDs must be equipped to meet social needs" (1971, p. 39). NSF Director, W.D. McElroy made more specific proposals:

there must also be a heightened awareness of the requirements placed on all science, and for this reason a significant share of the total resources available to NSF in the future must be devoted to the social and technological needs of the nation. . . . To bring the best of science to bear on the social and technological problems of society requires at least three steps. A larger number of the most creative members of the scientific community must be encouraged to associate themselves with the great problems of man and society; for even though not all of the world's ills have a scientific or technological base, the thought patterns of science and its intellectual-material accomplishments are proof that science has much to offer society (McElroy 1972, p. 389).

This acceptance of the new limits of discourse by top scientists, mainly those at the NSF, did not happen without a struggle. Defenders of basic science battled for budgets and resources against the proponents of a science that would meet national needs, forcing the issue as to whether the same basic science that saved the nation in the 1960s was enough to save the nation of the 1970s.

## The New Image of America Enters Policymaking

Since 1965, the NSF received pressure from the Subcommittee on Science, Research and Development of the U.S. House Committee on Science and Astronautics to adjust its national mission to meet growing social needs. As the Subcommittee's Chairman, Emilio Daddario (D-Conn) said in 1966 during a review of NSF's first fifteen years:

> within the present scientific, political and social context, NSF is operating in a manner which was satisfactory a decade ago but which does not appear adequate for either today or tomorrow. . . . Now it is time to make the Foundation into a broader instrument for forging and shaping the national policies to foster the national resources for science and to focus and direct them toward the attainment of great national goals: full employment, a clean world to live in, a population in balance with resources . . . (U.S. House Committee on Science and Astronautics 1966, p. ix).

As we have seen, powerful actors do not, by themselves, make policy. They must speak within the limits of what is sayable. In the mid-1960s, the U.S. was still engaged in a science race with the Soviets. The government still endorsed science as "a way of life,", and the NSF was funding quality science education to meet the Soviet challenge. Demands by Daddario began what became to be known as the Daddario-Kennedy Amendment, which then resulted in the NSF Act of 1968. The Amendment brought about organizational changes by restructuring the discipline-oriented directorates into area-oriented ones. The NSF's new areas of responsibility became basic research, manpower and education, institutional development, and science information. Despite these changes, the Act of 1968 did not challenge basic research as the main category for NSF funding.[2] However in the 1970s, with a new image of the nation challenging policymakers to position social and environmental problems within the NSF's mission, the status of basic science began to be threatened.

Speaking within the limits of what was now sayable, powerful actors in congressional committees with jurisdiction over the NSF articulated this notion of a nation in turmoil which could be saved by a "different kind" of science. Senator Edward Kennedy (D-Mass), Chairman of the Special Subcommittee on NSF of the Senate's Committee on Labor and Public Welfare, in the opening statement of the 1971 NSF authorization hearings recognized that

> We are meeting at a crucial period in the history of American science. Science and scientists are facing both unprecedented criticism and unprecedented opportunity. The critics contend that our scientific resources are too heavily concentrated in the defense area and that our scientific technology has too often ignored the needs of our environment. But even these critics must recognize how much we require science's help if we are now to cure our domestic ills, including pollution. That is the opportunity facing science today (U.S. Senate Committee on Labor and Public Welfare 1970, p. 1).

Also within the sayable limits, policy makers located a special kind of science education and research at the NSF, hence proposing the NSF as an institutional solution for the new national problems. Having watched the NSF work as a responsive policy instrument that produced basic research and quality scientists for the "Age of Science," policymakers like Kennedy argued for making the NSF into a policy instrument for new national needs. Challenged by the new image of nation in turmoil by domestic problems, Kennedy contested the NSF's appropriateness. As he put it:

> We live in an age of immediacy. From TV screens to the youth culture, the emphasis is on now. Each evening we view the human carnage which took place in Vietnam that morning and we are immersed in the immediacy of our problems —from the war to the economy, crime to pollution. . . . To solve these problems and build the kind of world we want, we need the best knowledge we can develop and the best educational system we can shape. The National Science Foundation has a unique role in meeting these goals . . . (U.S. Senate Committee on Labor and Public Welfare 1972, p. 1).

Sen. Kennedy continued pressuring the NSF to change, not only by bringing applied science and social sciences to the forefront of the NSF's mission but proposing an entirely new educational mission for NSF:

> The NSF has a major role to play in enabling scientists to meet the challenge of the seventies. Through its support for basic research and science education, the NSF prepares the way for important discoveries. Through its support for applied research and the social sciences, it can help mission-oriented agencies at every level of government utilize this discoveries effectively. The administration recognizes the significance of the NSF's role and has increased its authorization in 1971 (Ibid.).

Congress listened to Kennedy and appropriated $34 million (7% of NSF's total budget) for FY 1971 in order to establish the IRPOS's successor under the title Research Applied to National Needs (RANN). (see figure 3.1)

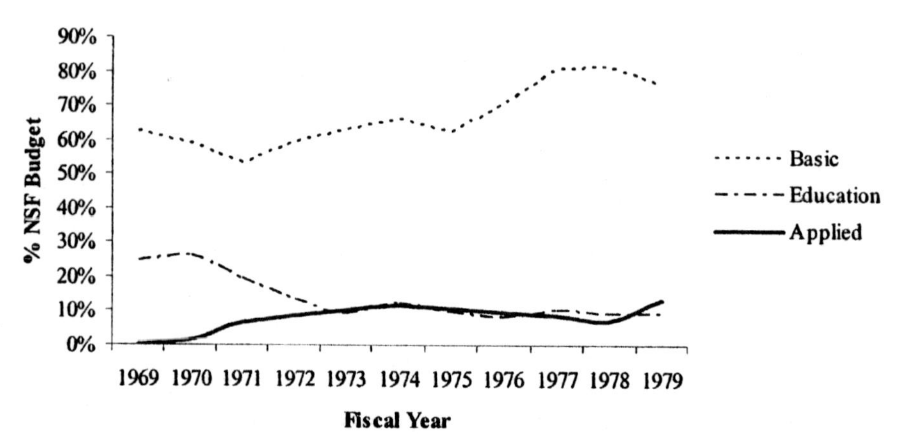


**Figure 3.1. Basic Research, Applied Research, and Science Education as Percentages of NSF's Total Budget: 1970s.**
Source: NSF's Annual Reports from 1969 to 1980.

While the NSF's national mission in the 1960s fit the desires of scientific academism, its emerging mission to address the new needs of the nation bothered them. As Sen. Kennedy, from his powerful position as Chairman of the Special Subcommittee on NSF, pressured the agency to deal more with national problems, scientific academism in NSF top positions resisted. Both Philip Handler, NSB Chairman, and William McElroy, NSF Director, opposed attempts to include applied science within the domains of NSF. As Handler put it before the Special Subcommittee on NSF of the U.S. Senate:

> One should not warp the scientific endeavor by converting all of our capabilities into applied activities and forsaking long range scientific ventures. . . . We must get on with the fundamentals today so that we will have the understanding we will surely require in a rather long distance future. . . . There must be an amply funded Federal agency which has the broadest possible license to support science, without concern for what the ultimate applications will be. That agency is the National Science Foundation(U.S. Senate Committee on Labor and Public Welfare 1970, pp. 7–9).

Following this line of resistance into the NSF's 1970 Annual Report, NSF Director William McElroy acknowledged that the "underlying new policies and programs for the 1970s" included a reexamination of science, engineering and technology "as basic tools of service to society" but that "progress in science cannot continue if its foundations—fundamental research—are weakened" (National Science Foundation 1970a, p. 1).[3]

## CAN 1960S QUALITY SCIENCE EDUCATION AND SCIENTISTS SAVE THE NATION?

When the NSB declared that the NSF's educational philosophy for the 1970s was "to educate scientists who will be at home in society and . . . a society that will be at home with science," this opened the door to multiple interpretations and struggles to define "being at home in society." The making of scientists and engineers at the NSF became a struggle between two competing interpretations. Scientific academists and NSF top officials endorsed a version that called for a moratorium on all educational and manpower programs except those aimed at producing top quality scientists for basic research. Senator Kennedy and science critics endorsed a version which called for education and training around national problems.

### Scientific Academism Defends Quality Science Education

In the 1960s, NSF officials sold science education programs with the promise of producing "a small cadre" of top quality individuals to save the nation from Soviet threat. In the 1970s, even with an image of nation redefined around domestic problems challenging them, NSF top officials resisted creating new science education programs, fearing they would further encroach upon not just applied science's budget, but upon the very image of "scientist" that they had created in the 1960s. Supported by official statements showing an oversupply of scientists and engineers, including statements by those in NSF's manpower unit, now the most trusted voice regarding the supply and demand of scientific manpower, NSF officials resisted both the creation of new education programs and the possibility of retraining the existing scientific workforce. As NSF Director William McElroy told Congress:

> It is the Administration's position that an adequate manpower supply exists in almost every field of science and engineering to meet Federal R&D program needs for the foreseeable future. . . . Very preliminary information on the employment situation for scientists and engineers, as appraised by the College Placement Council, the Office of Education, and the Department of Labor, indicates that the problem today is much more complex than one of reallocation of manpower. . . . Therefore, NSF believes that, when all of the complexities are taken into consideration, it would not be in the national interest for the Federal government to provide special Federal incentives at this time to encourage increased numbers of young people to pursue careers in science and engineering (U.S. Senate Committee on Labor and Public Welfare 1971, pp. 98–9).

Under the aegis of oversupply, McElroy, in order to differentiate between Ph.Ds working in basic science and non-PhD's working for space and military contractors, defined the problem of education in terms of numbers. Director McElroy continued:

> Current problems of oversupply in defense and aerospace industries involve, on close examination, mainly non-PhD scientists and engineers who have worked, over some years, on a single intensive technological aspect of a particular space or defense-related project. Completion or cancellation of these projects left these people in a particularly poor position to find new employment in an austere market (Ibid.).

Making clear that the non-Ph.D. ranks of applied scientists and engineers suffered from oversupply, McElroy protected scientific academism's pet program: the Fellowships. He assured Congress that the NSF intended to continue support of its graduate fellowships in the midst of oversupply:

> It would be a mistake, however, for NSF to discontinue all programs aimed at encouraging young people to pursue careers in science and engineering. There must be a bases support program sufficient to encourage outstanding young men and women who wish to pursue careers in science and engineering but low enough to prevent oversupply. Therefore, NSF is continuing to maintain a viable program of graduate fellowships based on a National competition. We believe that this program . . . will help provide the base of support required to insure at least a cadre of skilled scientific professionals over the long term (Ibid.).

This strategy proved successful until 1974, when fellowships for national needs and those for basic sciences received an equal share of NSF's education budget (see figure 3.2).

Another strategy that the NSF's administration used to resist the pressure to change its science education programs was to redefine social needs in terms of high-quality scientific education. As Lloyd G. Humphreys, Assistant Director for Education at NSF, put it to the U.S. House Subcommittee on Science, Research and Development in the 1972 NSF authorization hearings to determine priorities for science education: "The first and most important criterion is social need"(U.S. House Committee on Science and Astronautics 1971, p. 396). When asked by the Subcommittee's Chairman John W. Davis (D-Ga) to be specific on the meaning of "social need," Humphreys answered: "I mean the kind of education that our society needs to meet its commitments to our people, needs for scientific talent, needs for high-level technological talent, needs for teachers of science" (Ibid.). When asked again by Committee Chairman Davis if what he meant by social need was "purely science-oriented," Humphrey responded: "Yes sir, yes sir" (Ibid.). Following a logic sim-

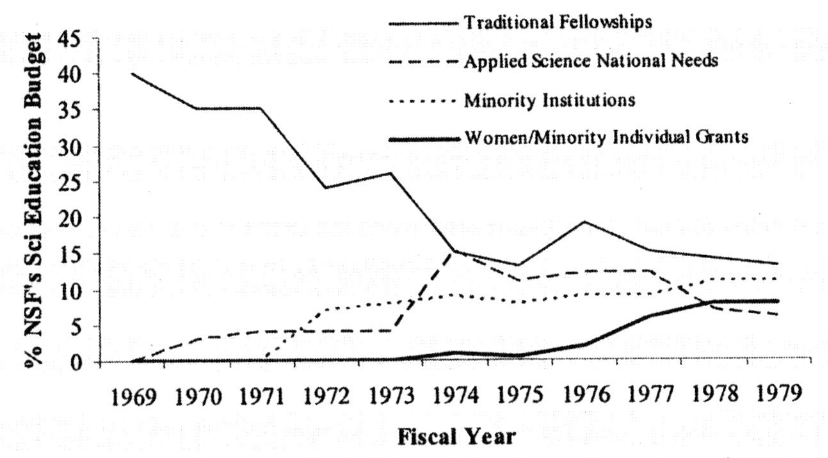


**Figure 3.2. Competing Categories in Science Education as Percentages of NSF's Total Education Budget: 1970s.**
Source: NSF's Annual Reports from 1969 to 1980.

ilar to that adopted by scientific academism ten years earlier, Humphrey claimed that continuing support for quality-oriented programs, such as the Fellowships and Teacher Institutes, best served society's needs.[4]

The NSF administration went beyond redefining social needs to protecting the Fellowships from needs-oriented programs. Under the pretext of oversupply, the NSF terminated the numbers-oriented programs initiated in the 1960s, such as the Traineeships, most of which went to engineers, and graduate support in the social sciences. With official knowledge from the NSF's manpower projections claiming the existence of "some serious potential [oversupply] imbalances for engineering and the social sciences," the NSF and the White House justified ending the numbers-oriented programs. This now presented the danger to the Fellowships of being redirected towards applied science for national needs. As Edward David, Science Policy Advisor to President Nixon, told Congress during NSF's 1972 Authorization hearings:

> We had in mind the decreasing size of the market for engineers and scientists. *This implies that our programs of educational support should be aimed at increasing quality rather than increasing quantity of graduates*. . . . Thus, we propose to discontinue NSF institutional programs which were primarily pointed at stimulating careers in science. Much of the reduction here results from a continuing phaseout of the *graduate student traineeships*. . . . That program was aimed largely at stimulating the massive increase in scientists and engineers needed for our space and defense efforts program in the 1960's. . . . At this time it is important to utilize scientists and engineers more effectively and to avoid overproduction . . . (Ibid., p. 653) (Italics emphasis).

Eventually, in 1973, NSF administrators completed the phase-out of the traineeships while continuing to grant 3–year Fellowships to ensure continued federal commitment to quality science education.

We have seen that the limits and forms of what is sayable at a given period include those "utterances that everyone recognize as valid, or debatable or definitely invalid" (Foucault [1968]1991). For how long could the NSF's administration keep making high-quality education, as conceptualized in the 1960s, part of the solution to national needs of the 1970s before it would be recognized as a debatable or an invalid solution given the new image of the American nation that challenged policy makers? Who ultimately recognized this, and how did they construct official knowledge in order to initiate acceptable education programs? This is where we now turn.

## Crisis Response vs. Sustained Responsibility Programs

In 1972, many factors contributed to pressure the NSF even further to change its attitude towards applied science for national needs. First, Nixon removed scientific academism from its powerful position inside the White House by abolishing the OSTP and the PSAC. Guyford Stever, the NSF's new Director, became Nixon's informal science advisor. Not only did Stever replace the old guard of elite presidential advisors, but he helped move the NSF toward applied research for national needs. The media described his job as follows: "provide technical answers to down-to-earth problems in transportation, energy, productivity, and health—a far cry from NSF's basic research charter" (1973a). To help Stever in his role as advisor, Nixon established an external Science and Engineering Council chaired by William O. Baker of Bell Labs and Simon Ramo of TRW, Inc. While Baker continued promoting the interests of scientific academism, Ramo endorsed redirecting science education towards social needs. While the NSF's administrators used the authority of manpower studies to claim an oversupply, Ramo predicted a shortage of people trained in dealing with new social problems. As he wrote in a speech in 1972,

> Long before 1990 it will become apparent that we have a shortage of properly trained people, particularly the interdisciplinary, practical intellectual. We cannot suddenly turn a large fraction of our engineers into experts on the social side for the problems or opportunities. Because it will be the only way to get started to get the job done, we shall, for a decade or more, create "social technologists" (perhaps we should say "poly-socio-econo-politico-technologists") in the school of hard knocks. These will be key performers in applying science and technology fully to the needs of our society. They will become expert at doing so by pragmatic, day-to-day synthesizing of arts and disciplines and experience and

> motivations and human ingenuity. . . (quoted in U.S. House Committee on Science and Astronautics 1972, pp. 50–1).

Regardless of his highly influential position, Ramo's statements did not have as much legitimacy as those coming from the NSF's manpower studies unit, by now the nation's most legitimate source of official knowledge about science and engineering statistics. However, his statements reflected how engineers in key influential positions reacted to the new image of the nation by proposing that "social technologists" will apply science to solve national problems. A look at the NSF authorization hearings in 1973 shows how this new proposed solution made it to the higher circles of policymaking. As Chairman of Subcommittee on Science, Research, and Development John Davis (D-Ga) put it to the NSF staff:

> A great deal of thought should be given at NSF as to what to do with people who are skilled but are unable to find an application for their skills, or whose technology moves off and requires them to be retooled (U.S. House Committee on Science and Astronautics 1972, p. 35).

NSF top administrators began to yield to these requests that called for retraining existing scientists. NSB Chairman H.E. Carter, who became Coordinator of Interdisciplinary Programs at the University of Arizona during his NSB tenure, responded to this pressure by saying that NSF science education programs were beginning to adjust to the changing needs for scientific manpower. Retraining included making scientists and engineers able "to deal with economic, environmental and social problems and [with] the need to develop and apply technology to provide at lower cost and higher efficiency the energy, transportation, housing and other services" (Ibid., p. 51). Carter actually recognized that education, as conceptualized in the 1960s, was no longer adequate for the present needs:

> A little over a decade ago when this Nation faced a severe shortage of scientists in the traditional disciplines, we developed incentives and provided resources to encourage the growth of high quality doctoral programs. The need is now different. We are just beginning to develop curricula and organize programs to meet the need for broadly trained interdisciplinary scientists (Ibid.).

Just how strong was the NSF's commitment to orient its science education program toward national needs? The NSF had been responding to these pressures as early as 1970 by assigning only minimal support toward problem-related programs, such as the Advanced Science Education Program, which included a component for the "training of environmental problem-solvers" (National Science Foundation 1970a).[5] Although it would appear that NSF was giving in to the pressures for applied research, its budget history throughout the 1970s shows that applied-research graduate education never surpassed the already shrinking quality-oriented graduate fellowships.[6]

Why did NSF top administrators so strongly resist this seemingly insignificant encroachment of problem-oriented programs into basic research? In 1974, when programs for women and minorities also entered the NSF's education budget, scientific academism defined the struggle between basic and applied science as one between the "sustained responsibility" *vis-a-vis* "crisis response programs." They defined "sustained- responsibility" programs as those that produced continuously high-quality scientists and engineers in the basic sciences, regardless of immediate national crises. On the other hand, they defined "crisis-response" programs as "crash" programs designed to meet immediate needs, including those of a particular group, but bearing the potential to compromise quality in science education. During oversight hearings on NSF science education programs in 1975, NSF administrators deployed the help of organizations representing higher education to call for sustained responsibility programs. Testifying on behalf of the Associated Colleges of the Midwest and the Great Lakes Colleges Association during oversight hearings on NSF science education programs, Dr Lewis Salter depicted the decrease in science education budget, from 45% of the total NSF budget in FY1959 to 12% in FY 1974, as the result of "turning off the manpower spigot" that followed Sputnik. Ironically, Salter categorized post-Sputnik programs as "crisis response," blamed them for present oversupply problems, and opposed the creation of new "crisis response" programs such as those related to energy and to women and minorities in science education (U.S. House Committee on Science and Technology 1975b, pp. 178–9). Regardless of any crisis or manpower situation affecting the country, Salter called for the NSF to concentrate on "sustained-responsibility" programs to improve the quality of science. This is exactly what NSF administrators wanted academia to tell Congress. Through representatives of higher education like Salter, NSF top officials pushed for quality-oriented programs while resisting the inclusion of "crisis-response" programs such as energy-related programs and those for women and minorities.

## "Blacks in Science" to Solve New National Problems

Criticizing the massive spending of federal funds on Apollo, which had no impact on improving their representation in science education, minorities set the stage to claim more participation in applied science for national needs. They criticized a nation whose priorities were to go to the moon at the expense of its minority citizens by excluding them from technoscientific training and education in the 1960s. Around the mid-1970s, these criticisms made it into the halls of Congress and into official statements when advocates of minority groups claimed that their including them in science was not only a matter of reversing the damage done in the 1960s but in the 1970s, a matter of national survival.

Marginalized from science education efforts in the 1960s, Black Americans were the first minority voice to appear in congressional hearings with the help of Senator Edward Kennedy. Mark Miles Fisher, Executive Secretary of the National Association for Equal Opportunity in Higher Education, claimed that black people, with the NSF's help, could contribute to solving national problems because they had intrinsic knowledge of Black America, specifically the inner cities, which whites lacked. While unemployed white PhD's could be retrained through NSF programs, they would never understand the problems associated with Black America. Testifying to the Subcommittee in support of $5 million for Ethnic Minority College Projects to the in 1972, Miles said:

> To use the area of Research Applied to National Needs as an example, one might see this possibility of involvement by the traditionally black colleges, along the following lines: The national problems of the Inner City could be dealt with through the resources of these institutions. People from these colleges are knowledgeable of the black community and should play a bigger role in solving the problems of the inner city (U.S. Senate Committee on Labor and Public Welfare 1972, p. 360).

Pointing at the substantial differences between NSF funds allocated to white *vis-a-vis* black institutions in FY 1970 (only 0.88% of NSF budget for institutions went to black colleges and universities), Miles argued successfully for more NSF funds. From FY 1973 to FY 1980, minority institutions received around $5 million or more per year. Other minority advocates, like Miles, were able to make the mobilization of Black Americans towards solving national problems a desirable characteristic of the nation's "manpower" in science and engineering. By 1973, the NSF's Assistant Director for Education Keith R. Kelson defined minorities' mobility as "variety" and declared the goal of its science education program "to provide for the numbers, variety, and flexibility of scientific and technological manpower needed to meet the Nation's changing requirements"(U.S. Senate Committee on Labor and Public Welfare 1973, p. 35).

Despite their own recognition of the need for "numbers, variety and flexibility," the NSF's top administrators still resisted by recommending, as authorized under the law, that no fellowships should be funded on the basis of disadvantaged backgrounds. In their 1974 report entitled *Federal Policy*

*Alternatives toward Graduate Education*, the NSB rejected Title IX, Part B, of the Education Amendments of 1972, which authorized graduate fellowships for capable persons from disadvantaged backgrounds. The NSB also discouraged programs for graduate students based on financial need, which had been modeled on the undergraduate Basic Opportunity Grant. Fisher's strategy to counteract this resistance was to go after funds for institutional development instead of going after fellowship funds, the sacred ground of scientific academism. Funds for Historically Black Colleges and Universities (HBCUs), besides providing high visibility to congressional representatives who supported them, would provide the necessary infrastructure to guarantee a minimum supply of black students into graduate programs in science. This would serve two purposes. First, HBCUs would become mechanisms for distributive justice, which some defined as one of the main national problems, by providing opportunities for upward mobility. Second, HBCUs would educate those who would solve Black America's problems. This is how Miles positioned HBCUs:

> institutions [that] have enabled hundreds of thousands of students shackled by poverty and racism to break free . . . [as] an existing mechanism that can be improved and used to intensify the positive efforts to equalize opportunity . . . and to guarantee the supply of blacks trained in science, as well as give a pluralistic mix in terms of the needs of our society(U.S. Senate Committee on Labor and Public Welfare 1974, pp. 123–5).

By outlining how HBCUs contributed to solving important national problems such as racism and poverty, Miles made "blacks in science" a significant statistical category within the nation's needs. This strategy proved at least partially successful, helping to securing 8% of NSF's education budget for minority institutions in FY 1974. Through associations representing white academia, the NSF's top administrators still argued that minority programs were to be seen as "crisis-response" and hence detrimental to an already overloaded labor market. Also, according to the NSF's manpower projections, there were already plenty of non-minority Ph.Ds who could be retrained if necessary. Minorities argued that the NSF's claims of oversupply were biased towards whites. Requesting Kennedy's Subcommittee on NSF to authorize the establishment of minority centers for Graduate Education in Science and Engineering, William Jackson, of the National Organization for the Personal Achievement of Black Chemists and Chemical Engineers, said

> I have heard here today that we do not really need to increase the number of scientists and engineers, but that is a reflection of what one of my colleagues calls "monochromatic assumption." The assumption is that since the number of

> whites in science and engineering are saturated no new programs are needed to increase the number of minorities in science and engineering(U.S. Senate Committee on Labor and Public Welfare 1976, p. 342).

By criticizing the "monochromatic assumption" of NSF officials, Jackson also pointed at the assumptions built into the NSF's source of authority that made claims about the number of scientists and engineers: the NSF's manpower projection model.

## Remaking 'Manpower' Models for Domestic Problems

At a time when the nation appeared plagued by social and environmental problems, an oversupply of technoscientific labor opened the door to question how valid was the NSF's source of knowledge about U.S. manpower. As a renewed NSB called into question models from 1960s, an image of a nation that needed mobile scientists to solve social and environmental problems flowed into NSF projection models. With old-guard scientific academists retiring from the NSB,[7] and strong voices calling for interdisciplinarity and for the participation of women and minorities in science,[8] the NSB called for a workshop "to carry out a critical comparative study of existing manpower analyses and the assumptions that underlie them" (National Science Board 1974, p. iii). According to workshop participants, which included NSB members and manpower experts, new manpower projections needed to address new national problems. One of the panels, entitled "Changes in National Priorities, Manpower Projection Techniques, and Requirements for Scientists and Engineers," stated the following:

> Changes in national priorities as they are reflected in government programs in health, pollution abatement, energy resource development, urban redevelopment and other areas can have consequences similar to those that took place [in the 1960's] because of the shifts in R&D expenditures. In effect, the activities undertaken to implement national priorities set up a series of demands for manpower at different levels of skill and occupational specialization in the public sector and, frequently even more so, in the private sector of the economy . . . [hence] projections that seek to account for the anticipated consequences of the pursuit of national goals for scientific manpower utilization in the next five or ten years [should] refer to social rather than to market demand (Ibid., p. 79).

Workshop experts found that the fixed-coefficient projection models, like those used in the 1960s and into the 70s, were somewhat limited for the new domestic agenda. After criticizing the over-reliance upon projections for federal R&D expenditures, they issued the following warning about fixed-coefficient models: "There is therefore, some tendency for oversimplified forecasts to be translated rigidly into educational policy. Yet fixed coefficient results have been surprisingly correct on the broad scale considering their limitations" (Ibid., p. 7). Workshop participants also criticized market models because these did not account for important market controls that were considered important to address new national needs: "Market models are criticized on grounds that the U.S. economy does not work in a purely market sense; there are numerous controls and rigidities in the system . . . [such as] equal opportunity and affirmative action brought about by legislation and other forces [which] have affected and shaped a number of significant trends . . ." (Ibid., pp. 7–10). The participants also proposed a shift towards projection models that were capable of forecasting mobility between disciplines, between specialties and between problems of national importance. For example, the panel recommended that "particular attention should be paid to the question of the extent to which occupational utilization depends on educational specialization, i.e., of flexibility in the use of scientific personnel." It also called for "methodology to understand the need for scientists and engineers in emerging priority areas such as environment, food shortages, or areas which will be affected by legislation requiring affirmative action." Furthermore, it recommended that "educational alternatives be provided to the traditional highly specialized Ph.D. program to prepare students for a broader range of careers" (Ibid., recomm). Accordingly, the NSF modified the Register. Explaining its switch to a new system that accounted for the race and gender of scientists and engineers, the NSF's 1974 Annual Report stated, "the Manpower Characteristics System, which took the place of the National Register of Scientific and Technical Personnel, . . . showed that 6,000 of these doctoral scientists held post-doctoral appointments, 21,300 were women, and 15,200 were members of minority groups" (National Science Foundation 1974, p. 100). Also, to explain a new interest in scientists and engineers in areas of national interests, the NSF claimed that "one major new program in the manpower area was initiated by the Foundation as a direct result of the recently experienced energy crisis . . . . Consequently, the Foundation started a program of study and analysis of information related to current and prospective utilization of scientific and engineering manpower in energy-related activities"(Ibid.). Following this NSB workshop came a number of problem-specific manpower analyses. For example, the National Research Council conducted a study on *Manpower for Environmental Pollution Control* (National Research Council 1977) and one for *Manpower for Primary Health Care* (National Research Council 1978).

Very quickly, new projection models made it into congressional hearings. In a 1976 "NSF Posture Hearing," the NSF reluctantly acknowledged the new model: "Although there remains much uncertainty about the accuracy of such projections [NSB's], it is clear that major problems such as energy, environment, and food production will require large numbers of highly skilled, flexibly trained people in the years ahead"(U.S. House Committee on Science and Technology 1975a, p. 8). The NSB's new chairman, Norman Hackerman, and the NSF's Guyford Stever officially recognized to Congress that the NSF needed "to learn how to make these people a little more mobile, more willing to move from discipline to discipline or problem to problem . . ." (Ibid., p. 23).

After 1974, official knowledge about new national demands to solve national problems translated into actual programs. From FY 1973 through FY 1974, graduate education in applied research for environmental- and energy-related problems increased from 3% to an average of 12%. For the first time, funding in these areas was given under such titles as *Fellowships in Science Applied to Societal Problems* and *Energy-related Postdoctoral Fellowships*. Although minority leaders like Miles had been partially successful in positioning minority institutions as parts of the solution to new problems, for which they received 8% of the NSF's education budget, they still had to convince the NSF that minorities were individuals worthy of such awards as the Fellowships.

Congress, not fully understanding the complexity of projecting mobile manpower for nation's survival, still demanded specific numbers to act upon. As we saw in Chapter 2, manpower statistics had become a legitimate source of official knowledge from which to direct policy making. As U.S. House Rep. Kenneth Hechler (D-WVa), of the Committee on Science and Technology, put it to NSF

> Dr. Stever and Dr. Hackerman, I think you are both in a unique position to look into the future and assess and appraise where some of our real brainpower shortages are going to occur 10,15, 20 years hence. . . . Could you quickly put into your computer all the various factors such as student enrollments, university interest, problems of the nation, problems of the world, jobs, state of the economy, and other things, and would you care to sound any kind of alarm or blow any kind of whistle for us as to what we need to be doing in order to protect ourselves 10, 15, 20 years hence? (Ibid., p. 23).

We have seen how minority leaders like Miles successfully positioned minorities as constituents of this mobile "brainpower" thus making them into a significant category for NSF funding at the institutional level. What minorities needed now was a legitimate official statement about their underrepresentation in science and engineering in order to show how few they were and to show lawmakers like Hechler how much the government needed to do. In 1976, for the first time, minority groups began using statistics from the Scientific Manpower Commission which showed that of the 207,000 Ph.Ds awarded in science and engineering, only 0.8% were awarded to blacks, 0.6%

to Latinos, and 0.04% to Native Americans. These statistics became instrumental official statements in the creation of the Science and Technology Equal Opportunity Act of 1980, in which Congress directed the NSF to begin publishing data on women and minority representation in science and engineering. However, if by 1974 minorities had found a way into science through institutional support, in 1976 they still had to argue that, in order to solve existing national problems, being black, Hispanic or Native American were important characteristics for scientists or engineers at the time.

## WOMEN AND MINORITIES IN SCIENCE AND ENGINEERING

Facing resistance from NSB and NSF top administrators to fund individual traineeships or fellowships for minorities, minorities began to argue that individuals with different racial and ethnic backgrounds were good for science. Their argument for individual awards differed from their argument for institutional support. If successful, their arguments for individual awards would elevate minority individuals to the highest levels of recognition from the NSF and the scientific community. Arguments for institutional support had retained the stigma of merely being requests for federal support of educational environments in order to nurture underprivileged populations. If minorities wanted full recognition as respectable scientists, it was indispensable for them to go after the Fellowships. If they wanted their research about problems affecting their own communities to be recognized as valid scientific work, they needed Ph.Ds and postdocs. This is what Shirley Malcom and Janet Welsh Brown, of the Office of Opportunities in Science of the AAAS, tried to do as they articulated a language of diversity that went beyond equal opportunity. They claimed that access to science and technology was good for the betterment of minorities, but it was also good for science. Testifying to the Senate's Committee on Appropriations, Janet Welsh Brown, head of the Office of Opportunities in Science at AAAS and president of the Federation of Organization for Professional Women, defined the problem—too few women and minorities in the sciences—and its probable causes, including the NSF's role

> . . . in the great scientific progress of the sixties, women and minorities were left further and further behind as the production of white male PhD's multiplied tenfold. It is a little known fact that our Black population received higher percentage of the science PhD's at the beginning of the 1960's than at the end—after all those NSF fellowships had been given out. Federal programs, therefore, actually contributed in some respects to widening the gap between white males and women and minorities in the sciences (U.S. House Committee on Appropriations 1976, p. 1851).

After pointing out that "the nation pays a very high cost for this inequality," Brown defined exactly where the highest cost to the nation lay:

> The third cost is the cost to science itself. The so-called "search for truth" and the scientific method require that ideas and values and assumptions be subjected to constant challenge and examination. I believe that diversity of cultural and other backgrounds helps assure that challenge and that diversity is necessary for good science. There is much in our accepted and unquestioned knowledge that is based on false assumptions and imperfect of incomplete research. . . . My hypothesis is that if there had been a healthier mix of women and minorities in the world of scientific research, more of those values and assumptions would be questioned earlier (Ibid.).

Brown concluded her presentation with statistics from the Scientific Manpower Commission, which throughout 1975, Betty Vetter, the Commission's Executive Director, had disseminated in a series of editorials in *Science*, making them part of scientists' discussion about the problems of the nation (Vetter 1975a; Vetter 1975b). Among those listening to Brown's statements in the Appropriations Committee was U.S. Senator William Proxmire (D-Wisc) who, during 1975, criticized the NSF for wasteful expenditures: "My choice for the biggest waste of the taxpayer's money," he said in March 1975, "has to be the National Science Foundation." But now, impressed by Brown's argument and statistics, he told Brown and the Committee

> Your statistics are overpowering. . . . I am glad that you have indicated what role the NSF can play in changing this situation. As I said earlier, I have been hammering away at them [NSF]. I have been trying to get them to change their minority and sex policies but primarily just asking them to hire more women and minorities and put more emphasis on those groups. But you spell out in specific detail how they can do it . . . I will write to the head of NSF, call his attention to your testimony, and ask for some answers on the extent to which the agency feels they can put these program into effect promptly(U.S. House Committee on Appropriations 1976, p. 1864).

What is important here is not that Sen. Proxmire found ammunition to go after his favorite target, but that Janet Welsh Brown defined the problem for the NSF to solve: the underrepresentation of women and minorities in science.

## Impact Assessment to Keep Control of Science Education

Losing control over their role as definers of the problems to be solved in science education, NSF administrators tried to keep their control over defining the solutions. With enough political pressure coming from champions of the underrepresented such as Kennedy and Proxmire, the NSF finally acknowledged that in science education, representation of women and minorities in science careers was its number one priority. However the NSF, by using impact assessment modeling (see figure 3.3), a problem-solving technique widely used in the 1970s by policy experts, remained in control of its science education programs' subsequent stages, mainly problem assessment, definition of constituency and programs, and problem impact.

Here, after problems were defined by "academic community, Congress, NSB, Administration, and society," the NSF assessed these problems and determined which it could not address because they required a changed emphasis in existing programs, or because they required new programs entirely (Ibid.). For instance, the NSF claimed that "financial distress of many colleges and universities" was an example of a problem that it could not address. This, of course, excluded most minority institutions from the NSF's domain. Then the NSF determined the composition of its constituency as follows: students, faculty, materials, and institutions. Again, in a "taxonomy of the Academy" to determine which kind of institutions fell in their constituency, NSF did not recognize HBCUs or all-female colleges and universities as belonging to separate categories than research universities. This meant that the interests of the latter guided the definition of programs. The NSF's next step was to use its constituency in order define new programs accordingly. Here, the NSF only listed traditional graduate fellowships and energy-related fellowships as "student access" programs. Women and minorities were still not recognized as "student" constituencies. Under "faculty access" programs, the

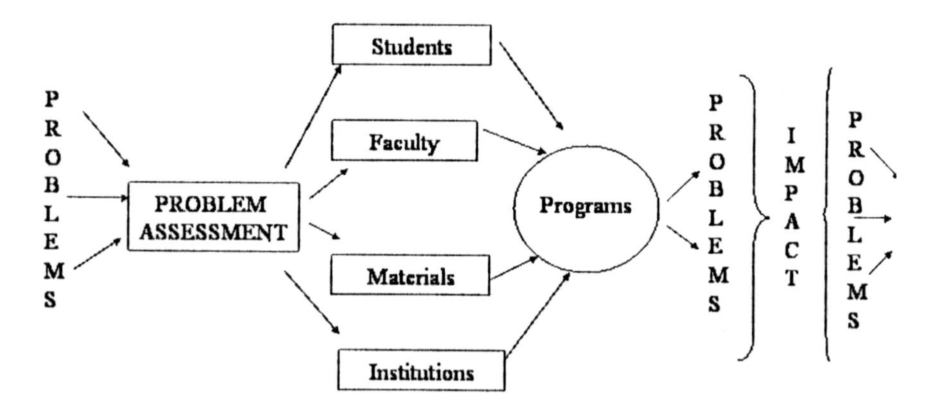


**Time Axis**

**Figure 3.3. Impact Assessment Model Used in NSF's Science Education Activities Cycle. Source: U.S. House Committee on Science and Technology 1975b.**

NSF created two categories: quality and quantity. While the former represented "intellectual capability," the latter represented numbers of women and minorities. Under "institutional access" programs, the NSF listed Minority Institutions Science Improvement. As we have seen, with only 8% of its education budget, this is the only significant program showing the NSF's commitment to minority representation. Although the NSF claimed that it used this methodology to do impact assessment of all its programs, it became a tool to screen and experiment on "crises-response" programs, such as those for underrepresented minorities. In the NSF's 1974 Annual Report, "Minorities and Women in Science Studies and Experimental Projects" were the only career-oriented programs to have experimental status with a combined budget of $0.6 million (National Science Foundation 1974, pp. 91–2). After "impact analysis," the FY 1975 budget for these programs had decreased to $0.3 million (National Science Foundation 1975). In short, the NSF's modeling of education meant that although images of the nation entered into the definition of the problem—through NSB's manpower recommendations, Janeth Brown's own definition, Kennedy's visions on NSF's role—, the NSF had maintained its control over defining the solution: its own programs.

Frustrated with their failures to increase their representation in science through NSF programs, minorities shifted their strategy to the newly proposed Department of Education (DoE). At the DoE, minorities saw the possibility to regain the power that NSF's modeling had taken away. As Jesse Jackson put it in his testimony to the Committee on Governmental Affairs of the U.S. Senate, assessing the transfer of the NSF's science education programs to the DoE:

> . . . by lifting education to cabinet status it would become more accountable to Congress, parents, teachers, students, local officials, private institutions and the press . . .The challenge of education today can only be met with significant public involvement and support, which requires strong leadership, a commitment to public participation in the highest levels of policymaking, and visible responsible administration . . . (U.S. Senate Committee on Governmental Affairs 1979, p. 13).

The Office of Technology Assessment (OTA), in charge of studying the impact of the DoE on federal science and technology activities, claimed that if its programs were transferred to DoE, the NSF would lose authority over science education. The OTA concurred with President Carter, who acknowledged that "consolidating those federal programs aimed specifically at improving access of minorities, women, and the handicapped [at DoE] will emphasize the administration's commitment to alleviating problems of inequity and discrimination in education" (Ibid., p. 264). In the end, the NSF only allowed the transfer of minority institutional programs to DoE, since those programs had the stigma of being "welfare programs" and the NSF did not want them under its jurisdiction. On the other hand, the NSF resisted transferring non-institutional programs for women and minorities to the DoE. This resistance, and Kennedy's commitment to keep women and minorities in science programs at the NSF or to make John Slaughter the NSF's first black Director in 1980, are also the result of the NSF's new national mission to solve social problems through applied science. With white males not being mobile enough to deal with all national needs, particularly those of women and minorities, the NSF wanted to remain in control of making mobile and interdisciplinary scientists out of different races, ethnic backgrounds and genders to solve national problems. This authority was finally granted by the Science and Technology Equal Opportunity Act of 1980 (PL-96-516) which Congress justified, arguing " it is in the best national interest to promote the full use of human resources in science and technology and to insure the full development and use of the scientific talent and technical skills of men and women, equally, of all ethnic, racial, and economic backgrounds" (Ibid.). In its own terms, and after a decade of power struggles with the NSF, Congress had incorporated into the making of scientists and engineers an image of a nation plagued with social and environmental problems.

## CONCLUSION

At the beginning of the 1970s, in the midst of both the triumphs of technoscience and the fears it bred, an image of the American nation at war with itself emerged. While most Americans celebrated the Apollo moon landing, some Americans, including a few scientists, were skeptical of this accomplishment and questioned science and technology's competence to solve domestic problems. In contrast with the 1960s, when relatively few scientific academists proposed basic scientific research and education as the solutions to the Soviet threat, in the 1970s, a bevy of science enthusiasts rushed to gain control of defining domestic national problems in a way that caused "a different kind of science" (i.e., applied science) to emerge as the solution. Throughout the 1970s, a renewed scientific academism at the NSF resisted this encroachment of applied science into the bastions of basic research, including the NSF's education programs.

Given the nature of the domestic problems, which included urban and environmental problems affecting minority populations, underrepresented minorities in science began arguing that white males alone could not solve the nation's problems. These claims only began to have a significant effect when the NSF's source of certified knowledge about scientific manpower came under attack by experts who realized that models from the 1960s did not reflect an image of nation with new domestic problems. Turning "mobility" into a desirable trait for the new workforce, the new projection models made "women and minorities in science" a significant category for the solution of new national problems, and hence worthy of federal support. Once minority representatives presented their own certified data that showed their small representation in scientific fields, Congress passed legislation to promote the representation of women and minorities in science as being "in the best national interest."

The government's creation of social-problem solvers allowed minority groups to constitute themselves as a category of statistical significance. By studying environmental and urban problems that were both in the national interest and in their own interests, some minority scientists not only served the nation but gave voice to other minority communities. The image of "scientist" that NSF helped create in the 1960s, although still dominant, had lost its hegemonic status. Even in a small way, minority scientists showed that a black person trained in specialized knowledge could also help save the nation from itself.

## NOTES

1. Motto on the cover of National Science Foundation 1970b.

2. For example, in 1969 when Congress appropriated only $6 million (1% of the NSF's budget) for Interdisciplinary Research Relevant to Problems of Our Society Program (IRRPOS), basic science took 82% of the NSF's budget, only 3% less than the budget a year after Sputnik.

3. NSF's budget reflected this tension between basic and applied research. As basic science's share of the NSF's budget decreased by 6% in FY 1971, applied science's share increased by the same amount.

4. For FY 1972, this translated into $22 million for Fellowships and Traineeships and $21 million for teachers' programs. In relative terms, 50% of the NSF's shrinking budget for science education was being devoted to quality-oriented programs while only 3% was spent on needs-oriented programs.

5. Having only $2.3 million in FY 1970 (or 1.9% of NSF's education budget) for graduate education programs oriented towards national needs, by FY 1974, the NSF had increased this kind of support to $12 million (or 15% of NSF's education budget) matching its support for traditional Fellowships.

6. Applied research had never exceeded 12% of the NSF's total budget. And basic science had actually grown from 59% in 1970 to 81% in 1978.

7. For example, terms expired in May 1974 for the following NSB members: Harvey Brooks, Professor of Applied Physics and Dean of Engineering and Applied

Physics, Harvard University; William Fowler, Professor of Physics, Cal Tech; Norman Hackerman, President, Rice University.

8. For example, the NSB Chairman was H.B. Carter, who was Coordinator of Interdisciplinary Programs at the University of Arizona. Anna Harrison, Professor of Chemistry at Mount Holyoke College, became the strongest voice for the representation of women and minorities.

## *Chapter Four*

# Japanese Technology Threatens America: Making Scientists and Engineers for Economic Competitiveness

> . . . as a nation, we must develop a consensus that industrial competitiveness is crucial to our social and economic well-being. Such a consensus will require a shift in public attitudes about national priorities, as well as changes in public perceptions about the nature of our economic malaise (Business-Higher Education Forum 1983, p. iii).

While the 1970s was a decade of policies and programs for domestic problems both social and environmental, the 1980s proved to be a decade marked with fiscal reorganization of the federal government and expansion of the R&D sector to promote economic growth. During this decade, the Soviet Union continued to pose a military threat, re-igniting old fears of communism. This double threat shaped an image of the American nation in the early 1980s. This image emphasized both national security against the USSR and economic competitiveness with Japan, now emerging as the nation's major perceived economic threat. In the late 1980s, to fight this double threat, the NSF became instrumental by developing a discourse that called for large numbers of scientists and engineers for both economic competitiveness and national security. In this chapter, I explore the emergence of the new image and then look at how actors at the NSF constructed a discourse that called for large numbers of highly skilled scientists and engineers to save the nation. Here I focus on the construction of knowledge around policy issues involving human resources for technological innovation in the 1980s. I am particularly interested in how government officials shifted policy and programs in education and human resources at the NSF, from those of the 1970s to new programs that had been redefined in terms of economic competitiveness.

What Sputnik did for science and scientists, positioning them as saviors of a nation facing a Soviet threat, economic competitiveness did for technology and engineers in the 1980s and thereafter. In the 1980s, some of the power struggles around education and human resources at the NSF were struggles to demarcate technology and engineers from science and scientists in a nation that had grown accustomed to see the latter as the answer to all its problems, both foreign and domestic. To redefine the NSF's education programs around the need for technological innovation, a new cast of NSF administrators would need to prove that mainly engineers would put the nation ahead in the technological race. I analyze here how the NSF invented a very powerful model and metaphor of the educational system in science and engineering—the pipeline—to show the nation just how many engineers would be needed to survive during both economic and military competitiveness.

## THE EMERGING OF A NEW IMAGE: JAPANESE TECHNOLOGY THREATENS AMERICA

In 1980, a year of intense presidential campaigning, headline stories told the American public how the nation was struggling to stay ahead of emerging competitors in the scientific arena. One article, entitled, "Science: America's Struggle to Stay Ahead," reported how "U.S. science for the first time [was] being seriously challenged by its major foreign competitors," mainly the Soviet Union, West Germany, and in the third place, Japan (1980c, p. 52). The report blamed "shortsighted management that [was] concerned with putting its money into short-term applied-research projects," like those of the 1970s, rather than into long-term basic and industrial R&D (Ibid.). The presidential election also served as a forum in which to further expose and question not only America's scientific capability, but also its economic survival. Both candidates, Carter and Reagan, by linking their economic platforms to revitalize the nation's productivity to R&D, brought science and technology to center stage. While showing how, during the 1970s, America's productive machine had gone from net exporter to importer for the first time in its history, popular media compared both candidates' programs to show the American public how each candidate planned to deal with the productivity crisis. One article titled, "The Productivity Crisis—Can America Renew Its Economic Promise?" summed up the candidates' plans to revitalize productivity through R&D as follows: "Carter promotes new Federal programs to help business cope with competition and regulation—a smallish made-in-America version of the government-business partnership in productive Japan. Reagan's philosophy, on the contrary, stresses getting government out of business's way and enabling 'free market' forces to take care of the rest" (1980b).

These popular media reports of America's economic malaise reflected the emerging image about the nation that would influence all sorts of government policies throughout the 1980s. The story went as follows: since America had slipped into a productivity crisis beginning in 1971 due to government's focus on social and environmental agendas, America now had to invest in a new kind of science and technology to renew its economic growth and the promise to its citizens for higher standards of living. Fulfilling this promise by solving America's social and environmental problems was no longer possible. National fulfillment would now come by reversing the productivity slip and putting the U.S. on top as a global competitor. However, the government and some of its constituencies still had to answer several important questions: Which kinds of science and technology? Where should the emphasis be, in basic research or in technological innovation? How should the federal government participate? To a great extent, answers to these questions would depend heavily on defining who was perceived as the most threatening enemy: the USSR? Japan? Both?

During the 1980 elections, both Reagan and Carter focused on Japan as one of America's main competitors and each proposed different ways of dealing with it. The popular media followed, creating an image of America threatened by Japan's advances in manufacturing technology and capacity to put new products on the market. Most of these stories focused on Japan's incredible technological capability which almost surpassed that of the U.S. The media turned Japan into a riddle for America to solve. In short, it became an American obsession to answer the following question: What makes Japan so successful in technology? As one article asked, "What makes Japan work? What is the secret of Japan's phenomenal success in business and technology? This question is rapidly becoming the obsession with the rest of the world" (1981b, p. 66). In trying to articulate this question, numerous stories concurred on two main points: there is something unique about the Japanese government's involvement in its industry, and the Japanese workforce possesses something unique that the U.S. workforce lacks. Responding to the question about government involvement, a 40-page special issue on "Japan's strategy for the '80s" claimed that "Japanese government is still playing an important role in the implementation of industrial policy . . . [their] Ministry of International Trade and Industry [MITI] determines the industrial structure and tells every industry what it can and cannot do" (1981c, p. 40). Responding to the question "Has Japanese Innovation Replaced Good Old Yankee Ingenuity?" posed in a Motorola ad in the early 1980s, articles responded, "with people as her only resource, Japan is outsmarting the United States in high technology products" (1980a, p. 751). However, the question remained: "How does she do it?" (Ibid.). Fueled by the presidential campaign, the media and politicians began to define the limits of what was sayable about the nation. First, America's main problem was *productivity*. Second, its main competitor was *Japan*. Third, Japan's secret of success seemed to lie in its *government-industry* arrangement and in the *education* of its people. Fourth, in order to compete with Japan, the U.S. had to look into these unique Japanese traits and decide whether it was to copy, adapt or reject them.

This image of the American nation threatened by Japan began to appear simultaneously with the re-igniting of fears about the Cold War. The U.S. continued to see the Soviet Union as a major military threat. While, in 1977, the popular media was celebrating Sputnik's 20th anniversary with headlines like "Sputnik Plus 20: The U.S. on Top" (1977), by 1980 the same media was questioning America's ability to stay ahead of the Soviet Union. As one article summarized, "U.S. science for the first time is being seriously challenged by its major foreign competitors. Although comparisons are difficult to make, one commonly used yardstick—the amount of money spent on research and development as a percentage of GNP—now shows the Soviet Union far ahead, with West Germany and the U.S. tied for second and Japan close behind" (1980c, p. 52). Still, in 1984, media headlines posed the question "Can U.S. Hold Its Lead over Soviets in Science Race?" (1984). The difference between this question and its twin, posed after Sputnik in 1957, was that now the race was for technology in general, both commercial and military. As one article concludes, "A nation that once intimidated America with its Sputnik satellite finds itself playing catch-up in its struggle for world dominance in technology" (Ibid., p. 51).

Needing public support for its military expenditures, the incoming Reagan administration fueled fears over Soviet technology and reignited the Cold War fears of the Soviet threat. Media stories endorsed the Reagan administration's "defense priority" of building "high technology weapons" (1981a). Policy experts from the Strategic Studies Center argued not only for another military buildup but also for placing more scientists and engineers in the new "science race" between the U.S. and the USSR (Ailes and Rushing 1982). Even the economic experts who built most of the foundations of Reaganomics defined the trade deficit with Japan as also being a national security problem with the Soviets, thus linking the productivity crisis with the Soviet threat. For example, Milton Friedman, leader of the Chicago School of Economics, and one of the main influences on Reagan's economic policy, was quoted in the popular media as saying: "if the decline in productivity were to continue for decades, the most likely outcome is that we would become a Russian satellite" (1980b).

The media, the presidential candidates, and the economic and policy experts alike, all having defined the main problem as that of productivity, now began defining the specific limits to this problem. Both Japan and the Soviet Union emerged as new and old enemies of the U.S. with their central governments orchestrating human resources for technology: in Japan's case, for economic leadership through manufacturing and productivity, and in the USSR's case, for military leadership through technological weapons. Now both government officials and technology experts were rushing to gain control over defining the solution. For example, deeply concerned about the nation's technological future in relationship with increasing Japanese competition, U.S. House Representative George Brown (D-Cal) of the House Committee on Science and Technology demanded, "What's wrong with us? . . . I'll tell you what's wrong. We don't know how to put our act together. We got to take hold and pull up our socks" (1981, p. 17). Experts on American technology, skeptical about using economic measures such as tax incentives and increased savings to solve the problem, proposed "a new spirit of American competitiveness, a rekindling of the desire to excel as a nation, to roll up the national sleeves and get to work" (Ibid.). Pointing out that federal funding for scientific education and R&D were what put America ahead in the 60s space race, these experts said, "What the nation needs is another Sputnik" (Ibid., p. 17). However, in contrast with the Sputnik of 1957, which had been visible in American skies, or compared with the civil unrest and environmental problems of the 1970s, which were visible on prime time TV, the "Sputnik of the Eighties," as one government official called it (Jennings 1987), still had to be invented. By the end of the 1980s, after a series of attempted policies to reignite government's interest in education and training of the nation's technological workforce failed, the NSF did manage to invent the "Sputnik of the Eighties." A look at some of these attempts shows how certain groups appropriated the new image of the nation and further defined the problem and its solution on their own terms: mainly those of technology and engineers. However, at the end of the decade, only the NSF's solution arose as the one that remained within the limits of what was sayable and acceptable in the new image of America.

## TECHNOLOGY TO SAVE THE NATION

In the 1980s, a wide range of proposals appeared to create incentives for technological innovation.[1] The proposal to create the National Technology Foundation (NTF) serves as one of the best examples of how the new dominant image of the American came to play a significant role in forming technology policy in the early 1980s. This Foundation, first proposed in 1980 by the House Subcommittee on Science, Research and Technology led by George

Brown (D-Cal), was meant to be the federal government's instrument for technological innovation and development, including infrastructure and human resources. As such, it was meant to be an institutional answer to the problem of productivity. The NTF was to do for technology what the NSF did for science (U.S. House Committee on Science and Technology 1980).

The congressional hearings surrounding this legislative proposal exemplify forums in which powerful pro-technology actors appropriated the image of nation to construct a discourse about technology as national savior. These groups and actors came primarily from government, industry, and academia, and had one fundamental belief in common: technology was the answer to America's productivity problem. They came to define very specific solutions for saving the American nation through technology and technologists. Among these groups and actors, we had the Committee on Science and Technology chaired by U.S. Representative Don Fuqua (D-Fla), and particularly its Subcommittee on Science, Research and Technology chaired by George Brown; Lewis Branscomb, the Vice President and Chief Scientist of IBM and now Chairman of the NSB; and the engineering profession and academic engineers such as Myron Tribus and Herbert Hollomon of MIT. Branscomb came to symbolize the new cast of high-ranking science-and-technology policy officials that were replacing scientific academism in the federal government. Like other scientists and engineers that followed in his path from corporate America into policy positions, Branscomb had a high regard for technology, he occupied a high level position in corporate America, and had a similar attitudes towards the free market as the Reagan administration.

These pro-technology actors defined the problem of the American nation in terms of productivity and trade-deficit; the solution, in terms of technology and engineers, and they included all of these in the language of the NTF Act of 1980. First, they claimed that U.S. productivity was falling behind that of most industrialized nations: "the productivity and rate of innovation of many national industries are lagging compared with the historical patterns and with the performance of the same industries in other nations, and are not sufficient to provide for a healthy national economy" (Ibid., p. 4). Second, they claimed that, since the mid 1970s, the international balance of trade had been unfavorable to the U.S. and that productivity was to blame: "the international balance of trade has been unfavorable to the United States for several years, including unfavorable balances in some industries heavily dependent upon technology" (Ibid.). Third, they positioned technology as the solution to issues of both productivity and balance of trade: "the development of new technologies promises fuller national employment . . . new goods or services for the national welfare . . . existing goods and services at lower costs. Thus, new technologies are generally counter-inflationary, facilitate market penetration, improve the national balance of trade, and support the United States dollar in international monetary exchange" (Ibid.). Finally, they also proposed that engineers and other technologists be the professionals to carry out the development of new technology, and they blamed previous science policies for neglecting these people: "the Nation has not given adequate attention to its requirements for engineering and technology manpower, to the need for engineering and technical research, or to the accomplishments of its engineers and technicians" (Ibid.). The main message: technological innovation will save the nation by boosting its domestic industries and improving its international economy. At the same time, new technologies will benefit American workers and consumers. With this in mind, the NTF Act of 1980 aimed to create a government agency that would promote technological innovation and the education and training of a highly skilled technological workforce.

## Can the NSF Do the Job?

The proposal for the NTF questioned the NSF's ability to serve the nation's emerging needs in technology. There were some actors who believed that the United States needed an institution to promote cooperation among government, industry, and academia, similar to MITI in Japan and the Ministry of S&T in Germany, but these actors believed that the NSF was not the institution. According to Herbert Hollomon, Director of Center for Policy Alternatives at MIT,

> every competitor nation has a mechanism to support technology related to its competitive position except the U.S [but] NSF is not capable of supporting industry-related technology. . . . I think that business starts in the universities of America, in the engineering schools. . . . I would like to see the NTF be clearly devoted to activities that are about doing, not studying, but doing (U.S. House Committee on Science and Technology 1980, pp. 773–7).

Although most of these pro technology groups agreed about the need for an institutional mechanism like the NTF, each gave it a different meaning. Academic "technophiles" like Hollomon saw the NTF as an additional source of federal funding for technology. Businesses, large and small, supported it as long as it was conceived as a non-regulatory agency, as a facilitator between government, business, and academia, and as a sponsor of industrial R&D. The engineering profession, since it had occupied a secondary role relative to science at the NSF, saw the NTF as a vehicle to reaffirm its legitimacy and presence in the federal government. These groups, however, questioned whether a restructured NSF could accomplish the new task or whether a new institution like the NTF was even necessary. Skeptical of scientific academism's past attitude towards technology, academic technophiles claimed that the NSF could not be transformed into an institution for technological development. As Myron Tribus, Director of MIT's Center for Advanced Study in Engineering, put it: "no man nor women placed in charge of the NSF . . . can turn the place around . . . to support deployment of technology . . ." (Ibid., p. 570).

Skeptics of the sharp differentiation between basic science and technology feared that such a distinction would translate into new and incommensurable federal roles in basic science and technological innovation. Lewis Branscomb, as both a Physicist/Chief Scientist at IBM and a self-declared "technophile" (Branscomb 1995), opposed the separation of technology from basic sciences and argued successfully that "NSF [was] the correct institution to operate and implement programs in engineering and technology . . ."(Ibid.). From his position as NSB chairman, Branscomb became the main character behind aligning the NSF's commitment to basic sciences with a new commitment to technology for economic competitiveness. In 1984, Branscomb would help bring Erich Bloch, an electrical engineer and Vice-President for Technical Personnel Development at IBM, to materialize this agenda as the NSF's new Director.

After pro-technology groups defined the problem and began to set the limits of the solution, more groups from academia and small business joined in to propose solutions in terms of "engineering manpower." However, they disagreed significantly on how to go about producing the technological manpower to compete with both the Japanese and the Soviets. The proposal for the NTF having failed, there were still concerns as to whether the NSF could take responsibility for educating and training technological manpower. Some individuals like Myron Tribus believed that the NSF could not do it while others rallied behind the NSF. For example, Senator Paula Hawkins (R-Fla) of the Committee on Labor and Human Resources, argued that Japan had an incredible way "to make engineers out of the workers from the assembly line while the robots took over the floor. Now you just cannot do that to the average line man in the U.S., and I think it is something that we are now seeing that must be addressed in pre-college . . . The NSF, I think, has a great responsibility for focusing attention on the plight that we are in today" (U.S. Senate Committee on Labor and Human Resources 1982, p. 133). Still others believed in the need for a comprehensive national manpower policy. What remained debatable was the level of government involvement. An administration committed to fiscal reorganization and reducing the federal government imposed limits to this involvement. The Reagan administration still had to be convinced that more engineers creating new technologies meant productivity and economic growth. For now, in the early 80s, the Reagan administration was discarding proposals that called for the creation of additional federal bureaucracy to promote engineering manpower like the NTF. A former officer at the Scientific and Technical Personnel Studies Section of the NSF, who later became Executive Director of the Office of Scientific and Engineering Personnel at the NRC, remembers the different attitudes towards government involvement in technological innovation, including manpower, in the early 1980s:

> There was never any doubt, and there still is no doubt in my mind, that the issue of science and technology, as an important positive force in society, is a bipartisan issue. If there's any dispute at all about policy, it's a question of how far should the government support it, not whether the government should support, but how far should it go. . . . The democrats would more likely go that direction [creating an institution like Japan's MITI] and clearly that's what happened at NIST, a big increase in their budget. Republicans were more skeptical of that kind of approach saying that if there is any way of doing this, the market contains all the right incentives. Give it to the federal government and the bureaucrats and they will screw it up . . . (interview with author).

## Engineers for National Security and Economic Competitiveness

The U.S. House Committee on Science and Technology produced a series of proposals that called for a national policy to provide the nation with the appropriate supply of engineers for both industrial productivity and national security.[2] In one of the first Committee hearings to debate national policy, entitled "Engineering Manpower Concerns" (U.S. House Committee on Science and Technology 1981a), the engineering profession established a clear difference between its contribution and that of scientists. Engineering professionals, represented by Robert Frosch, President of the American Association of Engineering Societies (AAES), positioned engineers as the key agents for competitiveness. He argued that engineers translated knowledge into design and production: "without engineers functioning to make this kind of thing happen, we would be in a very difficult position to even run our society, much less improve its competitiveness . . .if there is a major change in a lot of areas of economic expenditure, there will be a change in the demand for engineers" (Ibid., pp. 7–8).

Even though the U.S. enjoyed superiority over Japan and other foreign competitors in basic academic research, engineers questioned whether this type of research by itself was enough to ensure a lead in market shares and an improvement in the trade deficit. The U.S. needed engineers to do this job. Jack Geils, Senior Executive of American Society of Engineering Education (ASEE), stated that:

> The key point, of course, is that these applications are the work of engineers, not scientists. The engineer applies the new knowledge created by scientists in a

> practical way via designs of new and improved structures, products and services. Thus as technology multiplies, increasing quantities of quality engineers are vital to the Nation's well-being . . .We are in a recognized productivity slump and in my opinion only the engineering effort can pull us out (Ibid., p. 22).

To further distinguish between scientists and engineers, representatives of the engineering profession created a sense of urgency: the nation had plenty of basic research scientists but not enough technological innovators. Addressing the Committee on Science and Technology of the U.S. House of Representatives, Robert Gaither, President of American Society of Mechanical Engineers (ASME), said:

> I feel a necessity to clear up a point here, because I see the words 'engineer' and 'scientist' being bandied back and forth as equivalents . . .there is not today a shortage of physicists, chemists, nor mathematicians . . . The use of these words scientist and engineer I think has to be very carefully done. I would commend you to be careful with that (Ibid., p. 61).

Military leaders, supporting the engineers' claim of shortages, established the need for engineers in national security, hence positioning them as worthwhile for the government. During one of these hearings, Gen. Robert T. Marsh, Commander, Air Force System Command, claimed that the USSR and Japan were surpassing the U.S. in producing engineers and that an increase in defense spending by the Reagan administration created an immediate demand for engineers that the present supply did not satisfy: "There simply aren't enough engineers to go around . . . the engineer shortage problem is real," he said, pointing to the need for a national policy to deal with this new national crisis:

> I see a parallel between today's situation and that caused by the launch of Sputnik in 1957. After that shocking event the entire country was motivated to catch up, culminating in a national commitment to put an American on the moon. I hope such a shock is not required again. But nothing short of the national commitment we made for the Apollo Moon program will do here if we are to maintain leadership (Ibid., p. 32).

By invoking the Apollo program, Marsh began defining the new problem for the federal government to solve: creating large numbers of engineers for national security. On the other hand, high-level representatives of industrial R&D, now in top government positions, added to the demands of national security the demands of industrial productivity. Lewis Branscomb argued that, in addition to large investments in strategic defense weapons, the President's commitment to industrial productivity would bring even a larger demand for engineers:

> The driving force behind our industrial performance is the economic environment. The President has placed highest priority on creating the incentives in the private sector for an aggressive commitment by industry to the creation of new jobs at home and economic competitiveness abroad. This economic plan, if successful, will create the potential for growth. It is going to take a sufficiency of well-trained, motivated engineers prepared to address the problems facing our country to achieve the benefits that we want from that economy . . . if we are going to beat our competitors in productivity, we need manufacturing engineers with production skills who know how to automate a plant and who know how to manage quality work (Ibid., p. 70).

Since, in the new image of the nation, the main problem was productivity and not research, Branscomb positioned the training of engineers, and not scientists, as the educational problem for the government to solve. Up to this point, the government had neglected undergraduate engineering education, focusing instead on pre-college education through DoE and on science education through the NSF. Aware of this problem, Lewis Branscomb put engineering education on top of the NSF's educational agenda:

> Let me note that in between pre-college education which some, including this Administration, think is not NSF's role and the research which everybody, including this administration, think is important and is NSF's role—there is the problem of engineering education. I personally think that it is a problem for which an unambiguous, strong, and early effort is really important . . . I say so because the Japanese are not beating us with research . . . they are not beating us with development . . . What they are good at is production, and U.S. engineering schools are not training experts at production(U.S. Senate Committee on Labor and Human Resources 1982, p. 46).

We see here how Branscomb tried to redefine the NSF's mission according to the new image of nation under threat from Japan. He tried to make the NSF into an instrument of national salvation by assigning it the role of "training experts of production" with the help of the military, the government/ industrial R&D sector, and the engineering profession. Up to this point, Branscomb was at least rhetorically successful in inserting engineering education into the government's agenda. However, he lacked the official knowledge that technologies of government, like the NSF's manpower projections, could provide. What he had done was to further define the limits of what was sayable; i.e., engineering education is a legitimate problem for the federal government. In 1982, the year of these hearings, the NSF's education budget was at an all-time low (2% of the total budget). Around 1987, with the advent of official knowledge in the form of the pipeline projections, the situation for education would change. But now there was pressure from many groups to establish a national policy in the midst of the uncertainty that comes when there is no official knowledge informing the policy process. While respected officials like Branscomb called for putting engineering education at the top of the government's agenda, the Reagan administration still believed that education was not the business of the federal government. Actually, in 1982–83 it came very close to dismantling the three-year-old DoE and terminating NSF's educational programs.

## The Image of Nation Challenges 1970s Manpower Models

Congressional leaders resisted the administration's position on education by holding hearings outside of Washington and giving local educational leaders, and industrial leaders from small high-tech firms, a chance to participate in defining the solutions. Congressional leaders sought support for a national policy that would produce the required manpower. Unlike the executive who was committed to free-market economics, lawmakers did not believe that the free market by itself could compensate for the expected demands on manpower. Hence, they formulated a national manpower policy with the federal government as its main actor. In the words of Frank Press, President of National Academy of Sciences and former Science and Technology Advisor for President Carter: "I believe that the Government has the responsibility to monitor the supply and demand of scientific and technical personnel both in terms of quantity and quality and to supplement the free market role when it is in the national interest"(U.S. House Committee on Science and Technology 1981a, p. 131). Witnesses who testified on behalf of colleges and universities tried to define solutions in their own terms by showing how their institutions, properly assisted by the federal government, could contribute to the production (supply) of the manpower necessary in order to remain competitive. In doing so, they aligned their institutional interests with those of the nation. Armed with these demands, legislators could now legitimate redirecting federal support to their local constituencies for, in the long run, educational institutions within their localities would be contributing to fulfill the new national needs. In this way, economic competitiveness linked the local with the national. Competitiveness became the language through which America began redefining its international struggle away from a political and military and to an economic idiom, transforming the definition of the nation from a site within which individual interests competed into that of a single economic actor maximizing a collective interest (Downey 1995a; Downey 1995b; Downey and Lucena 1996).

An exemplar of these forums was the series of hearings entitled "Forecasting Needs for the High Tech Industry" taken to MIT by U.S. Representative

Margaret Heckler (R-Mass). She represented one of those Republicans who endorsed a national policy about manpower if it would ultimately favor the competitive position of the U.S. When it came to manpower for competitiveness, a former director of OTA's Educating Scientists and Engineers project, said, "I'm not sure if policy intervention vis-à-vis relying on free market would distribute neatly between Democrats and Republicans." Heckler brought together Massachusetts leaders of government, industry and academia to express their concerns about engineering shortages in the state's high-tech industries. A concern also arose among these leaders about the lack of models that could provide knowledge about the quantitative status of engineering manpower. As put by Roger Wellington, President of Augat, Inc., board member of the Massachusetts High Technology Council, and Chairman of the New England Council of the American Electronic Association:

> The critical problem is that we do not have a deep understanding throughout this country of the high technology shortage of engineers and the negative impact that this will make on the economic and political leadership of the country . . . During 1970–71 engineers were laid off due to aerospace and defense programs cuts. It became a media cause celebre. . . The public's distorted view of the so-called engineering bust, along with the anti-technology fashion of the Vietnam period caused young people to avoid engineering careers for several years . . . The great expansion of the 1970's in the electronics industry occurred during a period of declining defense expenditures. The projections of the shortages of the eighties is largely without the needs of the defense industry which will be superimposed on these shortages (U.S. House Committee on Science and Technology 1981b, pp. 12–14).

It is clear from this quote that manpower projections for the demands of the double threat, high-tech and defense industries, did not yet exist. These calls defined the problem and solution one step further: first, there was no reliable projection model and second, the new model had to account for technological innovation and national security's new needs.

More specific proposals for a national manpower policy came from Chairman Don Fuqua of the House Committee on Science and Technology and Chairman Orin Hatch (R-Utah) of the Senate Committee on Education and Labor. Generally speaking, these bills proposed to establish a national policy to ensure an adequate supply of scientists and engineers to meet the nation's future needs by establishing a central policy body that would dictate government actions (see for example U.S. House Committee on Science and Technology 1982b; U.S. House Committee on Science and Technology 1982a). Democratic members of both committees believed that a crisis of this magnitude needed a government reaction similar to the enactment of the NDEA of 1958 in response to Sputnik. As a joint report from both the Senate Committee on Education and Labor and the House Committee on Science and Technology states:

> These problems threaten to compromise America's stature in the international marketplace, weaken our industrial base, and undermine our national defense . . . The crisis depicted . . . is not unlike the situation our country faced in 1957, when the Soviets' Sputnik launch brought to a head concerns over mathematics, science, foreign language and technical preparation. In that era, Congress enacted the National Defense Education Act of 1958 (U.S. House Committee on Education and Labor 1982, p. 2).

By invoking the new image of America, Branscomb was able to place engineering education in the government agenda. Wellington had been able to call for new projections, and Congress was now demanding a national policy as sweeping as that of the NDEA. These congressional hearings, in various local forums, exemplify how the image of a threatened nation to be saved by a competitive posture towards technology began to take shape, how it spread and was reified. By invoking this image, groups legitimized their calls for federal support of education and manpower initiatives.

If these kinds of education initiatives were so important for the nation, why did the administration cut NSF programs in science and engineering education? Referring to these cuts, Daniel Berg, Provost of Carnegie-Mellon University, put it during hearings on "Science and Engineering Education and Manpower":

> The thing that many of us find inconsistent is that the goals stated by the administration, namely increased defense, productivity, reindustrialization, all require a strong research science and technology background. It is a sad inconsistency that many of us are having trouble with (U.S. House Committee on Science and Technology 1982b, p. 57).

Maybe what Berg did not understand was that the legacy of those programs came from a image of nation in which limits were defined around increasing participation in science with the hope of solving social and environmental problems.

If Branscomb and other pro-technology groups had begun to establish the limits of what was sayable, the Administration would be the last to define the limits of appropriation. That is, the Reagan administration would come to define how the relationship between image of the nation (America threatened by Japan), discourse (competitiveness), actors (Branscomb and other technophiles) and the destined audience (future technological manpower) was to be institutionalized. The Reagan administration did not yet believe that the best vehicle to achieve new educational needs in technology was the NSF's education programs. So the question became how to redefine education between the nation's new needs and the limits imposed by the Reagan administration, i.e., to reduce federal bureaucracy, government spending and intervention in education.

## Reaganomics Redefines Education from Equal Opportunity to Economic Competitiveness

Both the National Technology Foundation and proposals for national policy were opposed by business executives who viewed them as attempts to enlarge the federal bureaucracy and centralize the planning of manpower. The Republican administration wanted to minimize government intervention and serve instead as a catalyst for industry and academia to form partnerships that would pay for the nation's educational needs. The administration believed that in a free-market economy, imbalances between supply and demand of the products in question (engineering students) ought to be handled by the suppliers (academia) and by the consumers (industry). Dr. Douglas Pewitt, Assistant Director for Science Policy at the White House Office of Science and Technology Policy opposed legislation that called for manpower policy based on claims of shortages of engineers stemming from 1970s policies, saying to Congress:

> we cannot centralize national manpower planning for the nation based on current assessment of future needs . . . centralized manpower planning in the Soviet Union has been a failure . . .We cannot talk about problems of shortages of engineers without addressing problems of math and science education, especially at the pre-college level . . .they were built on a false set of assumptions [that] our nation could afford policies that emphasize the distribution of our existing resources over the creation of new resources . . . (U.S. House Committee on Science and Technology 1982a, p. 212).

The administration's belief in creating partnerships between government, industry and universities to begin redefining the problem of education and manpower in technology was exemplified by the National Engineering Action Conference (NEAC),[3] an impressive national effort that brought together academia, industry, government, and the engineering profession to focus attention on a significant national problem: the retention of engineering faculty and the shortage of engineers pursuing doctoral degrees and entering the teaching ranks. The Administration wanted to address these problems without national policy but through partnerships such as those between the Business-Higher Education Forum and the NSF.

Forums like the NEAC served two important purposes. First, they served as stage for the administration to inform its intentions of stripping the legacy of equal opportunity and domestic national needs from 1970s education and human resources programs and to redefining them in terms of economic growth and fiscal austerity. Responding to concerns on budget cuts for NSF education programs, George Keyworth, director of the Office for Science and Technology Policy for Reagan, addressed the NEAC, stating that the Reagan administration "discontinued support for many science and engineering education programs, like those at NSF, because they were rooted in the 1960s. That was an era of rapid economic growth in which the nation concentrated on distributing benefits and broadening participation. But in the 1980s the economy is slipping so we need to focus on economic growth" (1982).

During the congressional hearings entitled "U.S. Science and Technology under Budget Stress", George Keyworth had previously outlined the administration's position on the NSF's education and manpower programs. First, the rationale for funding was to move away from criteria based on equal opportunity and toward criteria that emphasized excellence. In Keyworth's own words: "My views on how we should approach the task of selecting high quality science have not been a secret. They can be summarized in one word—discrimination . . . [where] the first criterion must be excellence . . . In scientific endeavors we should, above all, advocate an unabashed meritocracy" (U.S. House Committee on Science and Technology 1981c, p. 22). Second, in spite of the numerous claims and the emerging evidence of manpower shortages in technological fields, this problem was to be handled by the free market, that is by suppliers and consumers of manpower, with minimum government involvement on the supply-side. According to Keyworth, this problem "must and can be worked out by those who supply science and engineering manpower [universities] and those who utilize it [industry]" without government intervention (Ibid.). Third, the responsibilities for education in science and technology had to be given back to state and local governments. Keyworth argued that as far as elementary and secondary science and math education, "[it was] the responsibility of the schools themselves and of the communities that [ran] them" (Ibid., p. 27–30).

Within the administration's new limits, the most powerful sectors of U.S. business and academia promulgated the importance of engineering manpower and education for economic competitiveness and national security. For example, after the NEAC, the Business-Higher Education Forum[4] began a campaign to spread the gospel at a national level "that industrial competitiveness [was] crucial to our social and economic well-being" and that engineering manpower and education were the foundations of America's competitiveness (Business-Higher Education Forum 1982; Business-Higher Education Forum 1983). The Forum connected engineering and the nation's competitiveness as follows:

> The state of engineering education in the U.S. is deteriorating severely with serious consequences for the nation's key industries and defense in a competitive, dangerous world . . . . The past two years have seen a growing recognition of the critical role that the availability of qualified engineering manpower plays in maintaining or reestablishing the technological preeminence that has been the back bone of this nation's economic and social achievements and the basis for our national security (Business-Higher Education Forum 1982, p. 12).

The Forum also advocated a coordinated effort between the demand and the supply of highly skilled labor involving the federal government only minimally. From this perspective, government intervention was only plausible on the supply-side. This meant that federal activities were acceptable only to help supply the military and American industry with enough engineers. In an analysis of technology policy, Branscomb confirmed this claiming that the Reagan administration largely confined itself to supply-side activities, that is federal activities to create new technology and technologists which would contribute to both government missions (e.g. SDI) and to innovation and productivity within the private sector (Branscomb 1993).

In line with supply-side economics, the administration proposed new technologies of government to find out what kind of manpower it needed to help supply. To be certain of its actions, the new administration needed certified knowledge. As Pewitt told Congress, "I have asked NSF to review its S&E personnel data collection to determine what more information can and should be collected to illuminate these issues [shortages] . . . NSF will establish a special technical manpower advisory group to review manpower data collection and analysis" (U.S. House Committee on Science and Technology 1982b, p. 212). This call resulted in the NSF 1984 report titled *Projected response of science, engineering, and technical labor market to defense and non-defense needs: 1982–1987* which served as a preamble for the pipeline.

In 1983, the famous report *A Nation at Risk*, released by the President's Commission on Excellence in Education, caught the country by storm.[5] Invoking the new image of nation under threat, the report claimed that "our unchallenged preeminence in commerce, industry, science, and technological innovation is being overtaken by competitors throughout the world" (National Commission on Excellence in Education 1983, p. 5). To overcome this threat, the report proposed an educational reform given that "the educational foundations of our society are presently being eroded by a rising tide of mediocrity that threatens our very future as a Nation and a people" (Ibid.). Focusing mostly on strengthening high school education, the report made pre-college education at the federal government's responsibility. However, following its commitment to supply-side economics, the government made it clear that the only acceptable form of involvement in education would be helping the supply of students. Therefore by 1983, most of the limits of what was sayable and doable about education under the new model of nation had been set. First, all levels of education, including pre-college and undergraduate education, had become acceptable areas for federal government to be involved in. Second, given its contribution to the main national problem of productivity, engineering education, as never before, became a recognized area for the government involvement. Third, government involvement was to be kept to a minimum: only on the supply side. Fourth, to reduce government spending and to pay for the costs of new education initiatives, the government promoted partnerships between industry and academia. Within these four established limits, a centralized manpower policy became unacceptable. Hence, the Engineering and Science Manpower Act of 1982, or any of its successors that called for national manpower policy, never became law. Eventually, President Reagan signed into law the *Education for Economic Security Act of 1984* (PL 98-377) which complied with the policy of economic recovery through a balanced budget, the federal government's diminishing role in the business of science and engineering education and manpower, and incentives to form partnerships between industry and academia to share the new educational programs expenses. However, one question still remained: if the limits of what was acceptable in education were laid out, how was the government to implement these at the NSF, which had a strong tradition in graduate science education but not in undergraduate engineering education or in pre-college education? Having laid out the limits of the new discourse about the nation with respect to education, I now turn to the NSF's construction of official knowledge that would allow the federal government to implement this new kind of programs.

## THE REMAKING OF TECHNOLOGIES OF GOVERNMENT

At the beginning of the "Star Wars/Samurai Era" (Lepkowski 1983), the Reagan administration nominated and approved Edward Knapp, a particle physicist, as Director of the NSF. In contrast to John Slaughter, a minority advocate and Carter appointee, Knapp brought the NSF into ideological alignment with the Reagan administration's policies of fiscal responsibility, especially by cutting science and engineering education programs. As he told *Science* magazine, "We're going to look at how people spend money. It's one of the reasons I'm interested in the budget. That's one thing a director can get a handle on" (1983, p. 991).

Knapp, the last of the scientific academists to direct NSF who favored science over technology, stated,

> Scientists have a responsibility to be active and responsible on problems of national security. Science and national security in this country are completely in-

> tertwined. There's virtually no way we can have our security without involving science and scientists. And in a sense, every scientist who does research and who has graduate students is participating in building the infrastructure for national security (Lepkowski 1983).

At the same time that Knapp helped terminate education programs from the 1970s, he revitalized NSF fellowships.[6]

Although he was in line with the Administration and with the scientific community, Knapp did not fully understand the mounting pressures calling for the NSF to respond to economic competitiveness. While key actors, like the technophiles, were crying crisis in technological innovation for competitiveness, he put the "chemists in charge of NSF" and pushed Keyworth's agenda of excellence and meritocracy as criteria to support basic science (Lepkowski 1983). And in 1984, Knapp's failure to understand how to adapt the NSF to technology for competitiveness brought about his replacement by Erich Bloch.

When Erich Bloch became NSF Director in 1984, he came in with a mission: to transform the NSF in order to meet the U.S.'s needs in the technological marketplace and economic competitiveness through the development of technological innovation, infrastructure, and workforce. Appointed by the Reagan administration, Bloch also had the support of the House Committee on Science and Technology which, having failed at making a national manpower policy into law, saw the NSF as a plausible policy mechanism to respond to the emerging manpower shortage crisis and to appease mounting public concerns.

Since 1982, the discourse of technology to save the nation and engineers to boost productivity had spread into the congressional committees that had jurisdiction over the NSF. Without new knowledge from the NSF on the state of scientists and engineers, these committees developed a sense of urgency that, as we have seen, led to attempts to create a national policy. Also, this sense of urgency led the committees to call for the creation of knowledge about scientists and engineers. In this new era of economic competitiveness, constituents began asking their representatives for explanations and projections that would inform the American public about careers for their children. As Senator Paula Hawkings (R-Fla) of the Senate Committee on Labor and Human Resources, with jurisdiction over the NSF's legislation, put it to Congress:

> So many citizens are upset and write to us: Why were we caught short? Why weren't the projections made? Why wasn't this data collected? Why didn't we know ahead of time that our son should have been studying engineering instead of sociology? (U.S. Senate Committee on Labor and Human Resources 1982, p. 134).

In the early 1980s, the Subcommittee on Science, Research and Technology of the Committee on Science and Technology, trying to respond to these pressures at a time when the NSF was undergoing significant transformations and budget cuts, appointed its own studies on manpower to the Library of Congress(Cooper 1983). At the same time, and maybe recognizing the limitation of such studies, Senator Kennedy of the Committee on Labor and Education pressured the NSF to develop a framework for its projections. During fiscal year 1983, when the NSF's science education programs were abolished, and engineering education emerged as an important element for competitiveness and national security, Sen. Kennedy outlined new responsibilities for the NSF, which, during times of both military buildup and economic competitiveness, was to begin charting the unknown future of national manpower needs:

> The Japanese now have doubled the number of engineering graduates in the last 10 years. We have held about level . . . We see the movement of R&D in the military area that is again going to draw [engineers] from the civilian area. I think that what we need are some flow charts and flow lines of what the implications of this are going to be in terms of our economy, in terms of jobs, where we are going to be internationally over a period of time. This committee cannot do that. The only ones that can do it are the National Science Foundation and the basic administration . . . [G]iven all these considerations, there is a flow line that is taking place in our society, and I think there is an agency that has to awaken this country as to what our needs are going to be (U.S. Senate Committee on Labor and Human Resources 1982, p. 46).

Kennedy directed these responsibilities to the NSF through Lewis Branscomb, who as former Vice President of IBM, had successfully managed IBM's manpower during the computer boom of the late 1970s. Unlike the NSF, IBM did not have to shift its manpower projections from the 1970s to accommodate the needs of a nation plagued with domestic problems that required interdisciplinary scientists and engineers to solve them. IBM's projections were specific to the new scenarios of the emerging technological marketplace. Kennedy now asked Branscomb, as Chairman of the NSB, to do at NSF what he had done at IBM:

> We have to know what you [Branscomb and IBM] have been doing. You cannot tell me that IBM was not thinking 10, 15, 20 years down the line. You were, and that is part of the reason that they were so successful. I would like to see the NSF try to give us some guidance about 10 or 15 years down the line and let us make the choices . . . (Ibid.).

Kennedy's demands reflected the lack of reliable projection models for the new image of a nation under threat from Japan's technological success. In addition, there was no consensus among manpower experts on how to carry out the forecasting of scientific and engineering manpower for the needs of the technological marketplace (Vetter 1985). Congressional leaders and manpower experts called for a conceptual framework that would allow for making projections about the new national needs of economic competitiveness and national security. The projection models proposed during the mid-1970s, based on assumptions of interdisciplinarity and diversity for domestic national needs, were now obsolete. Betty Vetter, Executive Director of the Scientific Manpower Commission (now the Commission on Professionals in Science and Technology), a private, non-profit organization that tracks the supply of scientists and engineers in the U.S., argued for the need of a new mechanism with which to conceptualize the supply of scientists and engineers for the emerging technological marketplace. Having been in charge of the Commission since 1965, Vetter understood that, given the new national realities, there was no plausible answer to the questions of manpower supply. In her report *The Technological Marketplace: Supply and Demand for Scientists and Engineers*, Vetter asked:

> what is the total supply of scientists and engineers in the U.S.? This seemingly simple question, often asked of manpower specialists, is not simple at all, and the answer is uncertain . . . increasing or decreasing the supply of trained scientists and engineers requires a lead time of four to eight years (quoted in U.S. House Committee on Science and Technology 1985, p. 526).

Alan Fechter, who at the end of the 1970s directed the NSF's forecasting of supply, demand, and utilization of scientists and engineers, also understood the need for a new framework and new data that would allow the government to make policy:

> unfortunately, we are currently unable to forecast supply and demand that far ahead with sufficient precision to draw strong policy inferences . . . we need to substantially strengthen our capabilities to generate long range projections for planning purposes . . . we have to be thinking now what we want to see 10 and 15 years from now in terms of what is coming out of the pipeline with respect to science and engineering (U.S. House Committee on Science and Technology 1985, p. 43).

Throughout the 1970s and early 1980s, first-year students' intentions to major in science and engineering had been well documented(Astin 1987). The reluctance of white males to go into these fields became more apparent. Not only did the U.S. need a new projection framework, it needed another kind of population group to rely on for its emerging needs in science and engineering. Betty Vetter knew this and made a public demand during the July 1985 congressional hearings on Supply and Demand of Scientists and Engineers:

> Forecasts usually are wrong because too many other factors also affect the outcome . . . Another data need is for accurate and timely trend data on all segments of the education/utilization pipeline in science and engineering, with the data broken out by sex and minority status . . . most valuable data allow us to examine the outcomes rather than the barriers along that pipeline (U.S. House Committee on Science and Technology 1985, p. 526, 533).

Here she acknowledged the unreliability of forecasting models used in the past. Recognizing that the decreasing white male population in science and engineering would not meet the supply needs of the nations, she called for new data based on sex and race. This shift in focus, from the predominantly white-male population being the traditional source for careers in science and engineering to women and minorities, was significant. Instead of acknowledging women and minorities for their contributions in solving domestic problems, as had happened in the 1970s, experts began to recognize the potential numerical contributions to manpower of both women and minorities for competitiveness. Also, the shift from outcomes (output) to movement along the different segments of the pipeline (flow) was significant, setting the stage for recruitment and retention efforts along the entire educational system.

## Engineers Provide the Model to Save the Nation

The Committee on the Education and Utilization of the Engineer of the National Research Council (NRC) responded to these calls for forecasting methodology and data in a timely fashion.[7] In late 1985 and in 1986, the National Research Council began publishing a 9-volume collection on engineering education and practice in the U.S. that, with the endorsement of the NAS and the NAE, reaffirmed the importance of engineering and engineering education to maintain the nation's strategic and defensive strength and its economic competitiveness in the international marketplace. Formed by high-ranking managers of America's most important corporations and engineering deans of its most notable universities, the NRC's Committee viewed the report as a tool "to assist policy makers in government, industry and academia by treating the various elements of engineering as a cohesive whole" (U.S. House Committee on Science and Technology 1985). The NRC's committee framed the conceptualization of the engineering workforce for this double national agenda as a complex engineering problem to be solved within the limits set by the Administration's concerns for fiscal reorganization and economic growth: "both responsibilities depend on problem-solving approach that is the heart of engineering" (National Research Council 1985, p. 27). This complex problem's solution would serve as the basis for future national policymaking without the need for a centralized manpower policy:

> this constant flux in the engineering workforce makes it more difficult to characterize accurately the engineering profession, to determine with any certainty who is what, how many engineers there are, and where they work . . . a clear understanding of the profession is necessary as a basis for national policymaking, for fiscal and economic planning . . . (Ibid., pp. 33–34).

To solve the problem of technological workforce, the NRC's committee proposed a social engineering approach. Jerrier Haddad, chairman of the NRC committee, claimed that his committee had developed a comprehensive flow model with which to understand the complex and cohesive engineering community:

> Properly populated with consistent data, this model should prove useful in analyzing the engineering community and its essential elements. Unfortunately, existing data bases and their varying definitions and their varying survey methodologies are inconsistent with each other. And this has made it impossible to use the flow model to its fullest advantage. We strongly recommend that a major effort be undertaken to see how best to arrive at the commonality of definitions, survey methodology, and diagramming methodology between the various governmental and private agencies that collect, analyze and disseminate data on science and engineers (U.S. House Committee on Science and Technology 1985, p. 328).

The flow model proposed a framework for tracking past events and for quantifying and forecasting future needs (see figure 4.1 ). The standardized model would study the behavior of subsets of the community (e.g., women, minorities) while reducing ambiguity in dealing with the overall technical workforce. The panel "considered a schematic flow diagram to be essential in understanding the complexity and dynamics of the engineering community" (National Research Council 1986, p. 2).

This flow model, designed by engineers, to study and predict the behavior of subsets of the general workforce, highlights social engineering at its best. The model is based on the balance equation from traditional energy and material balances:

$$Q1 + \Sigma fi - \Sigma fo = Q2$$

where

$Q1$ = number of people in stock at the beginning of period,
$\Sigma fi$ = sum of flows into the stock,
$\Sigma fo$ = sum of flows out of the stock,
$Q2$ = number of people at end of period.

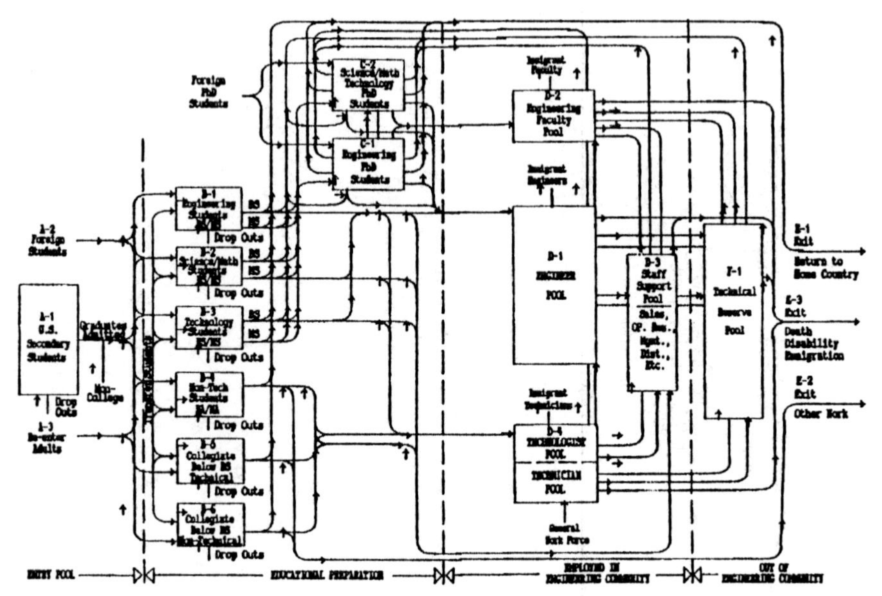


**Figure 4.1. Comprehensive Flow Diagram for the U.S. Engineering Community.**
Source: National Research Council 1986.

It was to "be adopted by the engineering community and used as basis and guideline for collecting data, analyzing flows and relationships, and projecting the effects of changes in flows and relationships" (National Research Council 1986, p. 7).

## Competitiveness and the NRC Model Come to the NSF

By 1984, the federal government, the business community, and corporate scientists like Branscomb and Bloch had appropriated the image of the nation and promoted a discourse in terms of competitiveness, its problems and its solution. Appointed in 1983, as recommended by the Business-Higher Education Forum, the President's Commission on Industrial Competitiveness released its report, *Global Competition: The New Reality*, which concluded that technology and human resources were the two most important elements to improve U.S. international competitiveness.[8] As the report stated, among the "factors that influence[d] our ability to compete . . . technology [was] our greatest advantage" (President's Commission on Industrial Competitiveness 1985). This claim, which received the approval of the President and the corporate community, established very specific limits of what was sayable and acceptable to do for the nation. Developing technology and human resources became the paramount solution to a very specific national problem: competitiveness. This new national agenda remained two-pronged—economic and military—until around 1987. When fears of the Soviet threat subdued, especially after 1986–87 when Mikaeil Gorvachev initiated his campaigns for *glasnost* and *perestroika*, the technological needs of economic competitiveness came first, before those of military buildup.

When Eric Bloch came to the NSF in 1984 from his position as Vice President of Technical Personnel Development at IBM, he put technology and human resources issues at the center of the NSF's agenda, redirecting the NSF's mission towards improving U.S. economic competitiveness.[9] Bloch thus proved Myran Tribus on MIT wrong when the latter said in 1980, "no man nor woman placed in charge of the NSF. . . can turn the place around . . . to support deployment of technology . . ." (U.S. House Committee on Science and Technology 1980, p. 570). Bloch, Chairman of NRC's Panel in Engineering Infrastructure Diagramming and Modeling, which developed the flow model, took this model to the NSF and made it the overall mission of the NSF's education and human resource programs. The NSF developed this model and made it into one of the most influential technologies of government: the science and engineering pipeline. As a high-level officer at NSF's Directorate for Engineering remembers:

> Erich brought the use [of the pipeline] and he was the first and only director that had this idea as a priority for NSF. He pushed the idea of the whole education system being important and pushed the idea of the importance of human resources, because before that, most people at NSF just focused on graduate school and research . . . Erich, being more forceful than most other directors, helped articulate the importance of education, competitiveness, and those who do not get a Ph.D. (interview with author).

When Bloch came to the NSF, its education budget had just descended to its all-time low: 3% of its total budget. When he finished his tenure in 1990, it had reached 13%, only to keep growing to 18% in 1996. This accomplishment was achieved by the science and engineering pipeline or the "Sputnik of the 80s."

## NSF and the New Technology of Government: The Pipeline

Erich Bloch became the first NSF Director to utilize the full powers of the NSF's legislative authority in the making of scientists and engineers for economic competitiveness. Authorized under the NSF Act of 1968, Bloch orchestrated data collection and analysis, the authority to recommend policies, and the institutional mechanisms to implement policies. He placed this orchestration under the roof of the Policy Research and Analysis Division

(PRA) which he reorganized as Director in 1984–5.[10] Bloch aligned the PRA with the Reagan administration's policies of orchestration and supply-side economics. Alzinga and Jamison have argued that the science policy agenda of the 1980s was marked by the strengthening of the corporate or economic culture and the weakening of the civic culture's influence. This was achieved in great part by "the policy of orchestration," in which government agencies "[sought] to shape consensus among representatives from the academic, economic, and bureaucratic policy cultures" (Alzinga 1995, p. 591). Erich Bloch achieved this style of orchestration through the PRA. In this division, policy experts recruited by Bloch developed a model of the educational system to shape a national consensus on the need for thousands of scientists and engineers to encourage economic competitiveness. In spite of significant criticisms from other sources of policy expertise like the OTA, the PRA's model achieved unparalleled national attention as well as support from the different groups that had, during the early 1980s, battled to define the course of manpower policy. The PRA's model achieved such attention because it embodied the image of the new nation and was developed within the limits of what was sayable and acceptable within the discourse of competitiveness. First, the presence of external threats created new economic and national security scenarios that demanded an increasing supply of scientists and engineers. The government had to help fill these manpower needs without intervening in the free-market, namely salaries. By shifting to demographics, the government could, without regulating salaries, create incentives to supply the required number of scientists and engineers for the new political and economic scenarios of "Star Wars" and for protectionist trade policies with Japan. The PRA pipeline studies did just that; it calculated the present and future supply of scientists and engineers, based purely on demographic trends without considering the effects of salaries, and it proposed intervention strategies in order to supply the required numbers. By looking at the assumptions that the PRA used to develop its model, one can see how it managed to stay within the limits of what was sayable and acceptable.

The first set of assumptions disregarded the effect of salaries in the demographic composition of the student population in science and engineering. These assumptions set the stage for government intervention in demographics without interfering with the free market. For example, the first assumption that the PRA made was that a fixed percentage of the student population chose science and engineering fields regardless of salaries: "irrespective of the many factors that enter[ed] into the choice of careers, natural sciences and engineering (NS&E) bachelor's degrees, exclusive of computer science, [had] been awarded annually to a relatively fixed fraction of the 22 year-old population for about three decades" (National Science Foundation 1990,

p. 189).[11] By looking at enrollment history, the PRA projected this fixed fraction to be 5% of the 22-year old population, irrespective of ongoing increments in college enrollments. The PRA claimed that "according to the Bureau of Census, the decline in the number of people in the 22-year-old age group [would] continue until after the mid-1990s" (Ibid., p. 192). Given the new national needs of manpower for productivity, this demographic picture framed the following question: how could this fraction be enlarged within the limits of what was sayable and acceptable?

In line with recommendations from manpower experts like Betty Vetter, the PRA focused on specific population groups that could be redirected into science and engineering. For example, the PRA claimed that "there [had] been a slow but persistent rise in the rate of conferral of baccalaureate degrees to women offset in large measure by a decline in the conferral rate to males. Most of this rise occurred between 1972 and 1982" (Ibid., p. 191).[12] The PRA also showed the potential contribution of underrepresented minorities to enlarge the 5% fraction, claiming that "although the participation rate of the underrepresented minorities [in the 5% conferral rate] [had] been substantially below average, blacks and hispanics [were] expected to increase their proportion of the total college age population from about 23 percent in 1989 to 28 percent or more in the year 2000, and 30 percent or more in 2010" (Ibid., p. 192). The PRA recognized women and minorities as untapped resources whose potential contribution was not being directed towards science and engineering but whose future contribution could help economic competitiveness.

The PRA drew the following conclusion from this demographic scenario: "without some positive action to substantially reverse the decline in student preferences for declaring NS&E majors, bachelor degree awards in these fields are likely to decline"(Ibid., p. 194).[13] Having removed salaries from the picture, the PRA, without intervening in the free market, paved the way for policy makers to increase the student population that went into science and engineering. The PRA's assumptions about the demand side also reflected the new image of nation. It claimed "because of the complexities in the utilization of NS&E training and limitations of occupational census data, quantitative projection of the demand for individuals with NS&E knowledge and training is highly uncertain, and was not attempted in this work. Instead, the average production during 1984–86 has been taken as a proxy for future demand (about 210,000 degrees per year). This proxy is conservative" (Ibid., p. 194). Not only did the PRA select the years with largest production for its base numbers regarding demand, but it dismissed any quantitative projection, such as salary or even government spending on R&D, which the NSF had included in previous fixed coefficient projections. The argument here was that the government had never embarked on large scale spending on the kind of

R&D that was now required. R&D spending during the Cold War, or for the national needs of the 1970s, could not be used under the present circumstances. On the demand side, the PRA also disregarded the effect of salaries, keeping away any possibility of government intervention in the free market. For example, the PRA claimed that "based on observations from the past decade, it appears unlikely that the labor market for NS&E bachelors will solve this emerging problem by steering a much larger fraction of undergraduates into NS&E majors, for two reasons: first, in the past there has been virtually no relationship between changes in the relative starting salaries and degree production in the combined NS&E fields . . ." (Ibid., p. 195).

Given these assumptions about supply and demand, the PRA concluded that if things were left as they were, "the cumulative shortfall of bachelors in the year 2006 would be about 675,000 with 275,000 being in engineering degrees" (Ibid., pp. 194–5).

After defining the problem as indicated above, the PRA also defined the solution to the upcoming manpower crisis through attraction and retention strategies in the form of programs coming from the NSF (see figure 4.2). According to the PRA,

> [t]he number of qualified personnel in NS&E fields can be increased through two strategies: attracting more new people into these occupations, or into the NS&E pipeline, and retaining a larger number of those who are in the NS&E pipeline (Ibid., p. 223).

## The Pipeline within NSF

Both inside and outside the NSF, the pipeline allowed policy experts to foresee possible futures and to inform policy. As Alzinga and Jamison argue, "foresight became one of the central policy methodologies [in the 1980s]. The idea was to bring various actors together, chosen on the basis of their specialized knowledge, to visualize possible futures and select particular technological options that then became inputs in policy. Foresight thus involves a strong element of social construction, whereby governments or agencies connected to government seek to shape consensus among representatives from the academic, economic, and bureaucratic policy cultures" (Alzinga 1995, p. 591). The pipeline became the NSF's "foresight" methodology for human resources, but before the pipeline could influence policy, Bloch and his PRA experts needed to establish consensus within NSF.[14] Once consensus was reached, the pipeline studies, with their assumptions, conclusions and recommendations, would become the official knowledge that, for almost a decade after 1985, guided the NSF's education and human resource programs. The search for consensus, however, did not occur

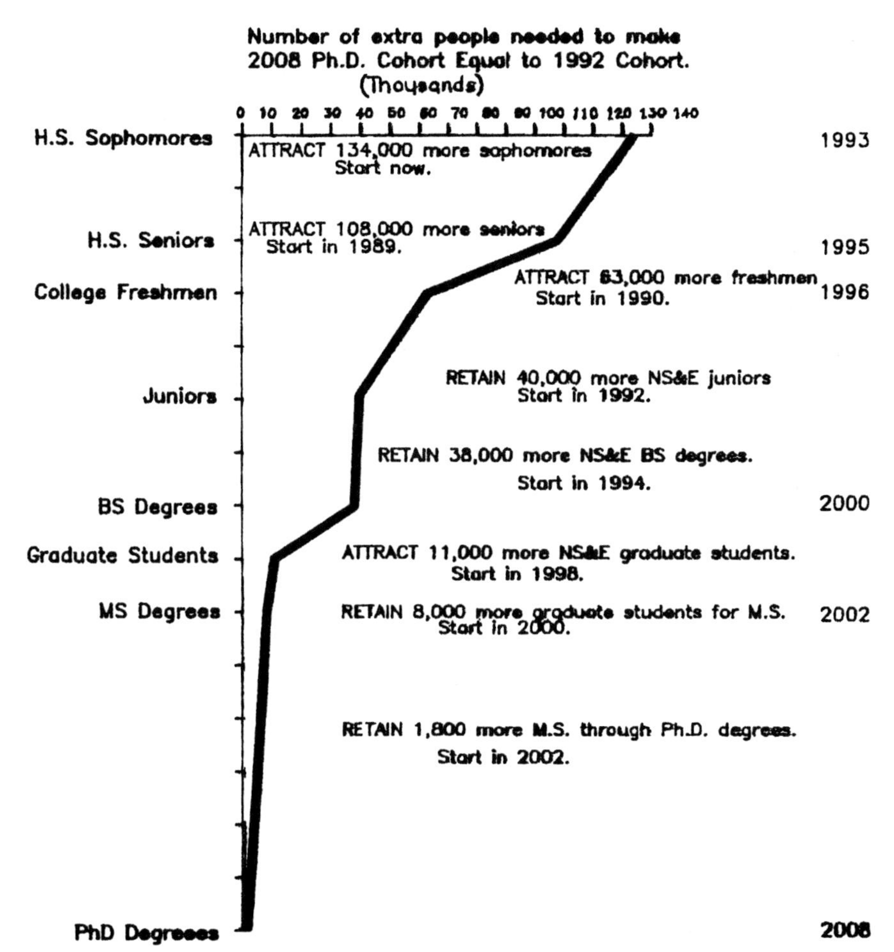


**Figure 4.2. The Science and Engineering Pipeline: Strategies for Reducing the Shortfall in Natural Sciences and Engineering.**
Source: National Science Foundation 1988c, p. 20.

without some resistance from other sources of policy expertise.[15] But the PRA had found a way to consolidate the legitimation of the pipeline within the NSF. First, there was disagreement among the science policy community about the best way to project numbers of scientists and engineers. While the PRA based its projections on demographics, the SRS and the OTA published reports showing the limitations of demographic-based projections. Both the SRS and the OTA advocated market-driven analysis. But Bloch's and the White House's powerful policy of orchestration which would supply the U.S. with a large number of scientists and engineers without intervening in market forces by fixing salaries, kept dissenters quiet until 1993.

As soon as Bloch became the NSF director, the PRA began a series of reports advising the administration about the demand-supply balance of scientists and engineers. These were followed in 1985 by briefings to the NSF and the NSB and by the modeling of discipline-specific pipelines for all NSF Directorates who embraced the pipeline.

The team put in charge of the PRA Division endorsed a policy practice known as Computer-Aided Science Policy and Research (CASPAR). PRA top officials argued that CASPAR would democratize the policy process: "The practice of data driven policy analysis is no longer the monopoly of the few who have access to or can program a mainframe, or who are privy to a sophisticated analytic technique, or possess a database."[16] Whether it democratized policymaking within the NSF or not will is the subject of another analysis. But in the late 1980s, through CASPAR, the PRA accomplished an unprecedented level of acceptance for the pipeline within NSF staff, from the Director and the NSB all the way down the organizational hierarchy. The PRA achieved this by disseminating the pipeline model throughout the NSF, using programmed spreadsheets that presented the PRA's assumptions as constants. Before briefings and board meetings, the PRA distributed diskettes to NSB members.[17] Program managers of the NSF's various divisions could now play with supply variables and create a variety of future scenarios specific to their disciplines and could make appropriate claims of manpower shortages.[18]

## The Pipeline Outside of the NSF

As soon as Bloch became Director, he began filing requests to the OSTP and the NAS for projections of scientists and engineers. Bloch's background at IBM, and at the NRC's Committee on Engineering Education and Practice, plus his close relationship with the Reagan administration, made him trustworthy in high-level policy circles. He lobbied for including the pipeline studies projections in the NSF's most consulted source of official knowledge, the *Science &Engineering Indicators*, which until then had relied on external contracts to develop market-based supply models for its projections (Dauffenbach 1983). For the first time, in its 1987 projections, the NSF included the pipeline projections as follows:

> In the 1990's, changing demographics may impose restraints on the supply of newly trained scientists and engineers . . . In fact, with the population of 18- to 24-year olds expected to decline by 23 percent between 1980 and 1995 college enrollments could decrease by as much as 12 to 16 percent by 1995 . . . These combined forces may hinder workforce ability to meet demand increases over the next decade (National Science Board 1987, p. 352).

Considering the respectability of the *S&E Indicators* in the U.S. government and its influence on all sorts of policy decisions, this is significant. For example, disseminating the pipeline outside the NSF in official literature like the *S&E Indicators* greatly influenced the new immigration policy, signed by President Bush, which gave special provisions to foreign nationals with technical and scientific expertise.

Bloch took the pipeline argument, now carved in stone and receiving the full support of the NSB, and tied it to the need to increase NSF funding. The PRA's recommended retention and attraction were built into the visual image of the pipeline to indicate the problem points and the proposed solutions to overcome the projected shortfall (see figure 4.2 above). This visual image became the NSF's poster child. The NSF's assistant directors, division directors, and program managers not only used the pipeline model to justify their own budgets internally, but also took the pipeline model on the road to obtain public and congressional support for their own programs. A high-level officer involved with NSF engineering education programs remembers the benefits of taking the pipeline image to the halls of Congress:

> Well, it's a very oversimplistic and not a very humanistic way to talk about students. It's a very mechanical model but that is important to some people. I've seen enough benefit from using the metaphor if you have to explain a member of Congress in 30 seconds what the problem is. This is something they can get and then they have benefited us in terms of getting support for the program. So it doesn't bother me all that much because I'm seeing the benefits (interview with author).

This is exactly what Erich Bloch did successfully when, in 1987, he took the pipeline message to Congress and successfully argued successfully to increase the participation of women and minorities in science and engineering as essential to economic competitiveness:

> . . . if we want to supply our industries and government and our universities with the human power that we need in the future . . . we need to concentrate on the groups which are underrepresented today in the scientific engineering areas—women and minorities. That is something that the Foundation has been focusing on and will have to focus on increasingly in the future (U.S. Congress Joint Economic Committee 1987, p. 9).

In the NSF's 1988 Annual Report, Bloch's statements reflected how much the pipeline had achieved for the NSF's education programs. As he wrote in the report's front-page statement,

> Early in 1987 the administration had asked Congress to double the NSF budget over the succeeding 5 years. This was unprecedented. It recognized

> the importance of science and engineering to economic competitiveness . . . our education programs are growing vigorously, with excellent support for the administration and both parties in Congress . . . Because demographic trends demand it, we will renew our efforts to attract larger numbers of women and minority-group individuals to science and engineering. Women and minorities must be recruited more successfully if we are to have the people we need in science and engineering. This is absolutely crucial(National Science Foundation 1988a, p. 1).

Since 1987, the NSF's education budget has grown continuously, not only in proportion to its total budget, from 6% in 1987 to 19% in 1996, but to unprecedented total levels, from $99 million in 1987 to $600 million in 1996 (see figure 4.3).

## The Pipeline Industry

With the help of the visual model of the pipeline, Bloch was so successful in delivering the pipeline message to Congress that his idea of creating the Task Force on Women, Minorities and the Handicapped in Science and Engineering was passed into law by Congress in 1987.[19] Congress established the Task Force to examine the current status of women, minorities, and the handicapped in science and engineering and to coordinate existing federal programs to promote their education and employment in these fields. The Task Force had the authority to "develop a long-range plan to advance opportuni-

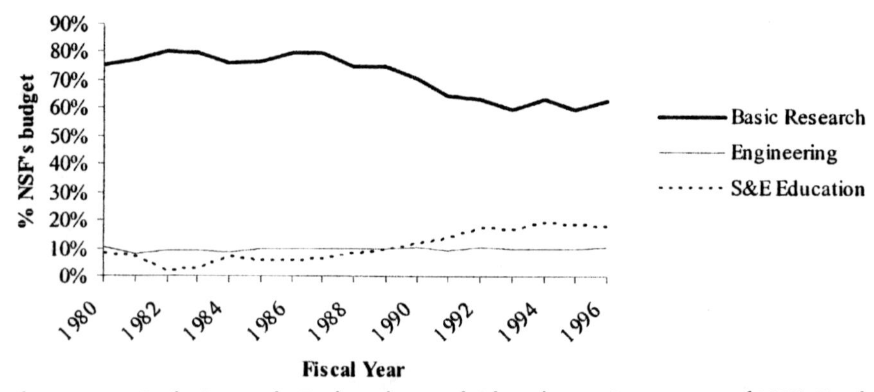


**Figure 4.3. Basic Research, Engineering, and Education as Percentages of NSF's Total Budget: 1980–1996.**

Source: NSF's Annual Reports from 1980 to 1997.

ties for women, minorities, and the handicapped in federal scientific and technological positions and in federally assisted research" (PL 99-383). Without precedent in the history of science education, the Task Force was co-chaired by a Hispanic and a woman.[20] It was directed by another woman, and comprised mainly of female and minority representatives from government, industry, and academia. The Task Force published the report *Changing America: The New Face of Science and Engineering,* including specific recommendations for many sectors of American society to implement its programs and policies.[21]

Invoking the image of nation under the threat of Japanese competitiveness, and using the pipeline as its foresight methodology, the Task Force identified specific groups (Blacks, Hispanics, American Indians, People with Disabilities, and White Women), it established their current participation in NS&E fields, and, by means of group-specific pipelines, it outlined group-specific strategies for attraction and retention. Furthermore, the Task Force placed the responsibility of increasing the representation of underrepresented groups in science and engineering upon "key players in American society." By making the representation of underrepresented groups in science and engineering everybody's business, and a "goal for the American nation", the Task Force opened the door for the pipeline industry.

One of the pipeline model's most important features was that it provided a foundation for longitudinal analyses of behavior of key subsets of the general population, i.e., women and minorities (see figure 4.4). It accomplished this by using desegregated flow diagrams, which focused on one particular subset of stock such as "black women in engineering" or "Hispanic men in physics." These group-specific models also made it possible to pinpoint "leaks" in the flow of a particular group at specific points in the educational pipeline. These "leaks" came to be known as "behaviors" that required institutional fixes, such as those by offices of recruitment and retention that had been established at colleges and universities and sponsored by the federal government and industry. As such, the pipeline became a technology of government that helped government and industry keep track of the science and engineering bound student population beginning as early as middle school. Also, it helped to incorporate women and minorities into the economy, making them social citizens once again, but now in accordance with the national context of the 1980s.

The Task Force recommendations established a new dimension in the relationship between the federal government and educational institutions, spanning from kindergarten to research universities. By recommending that the federal

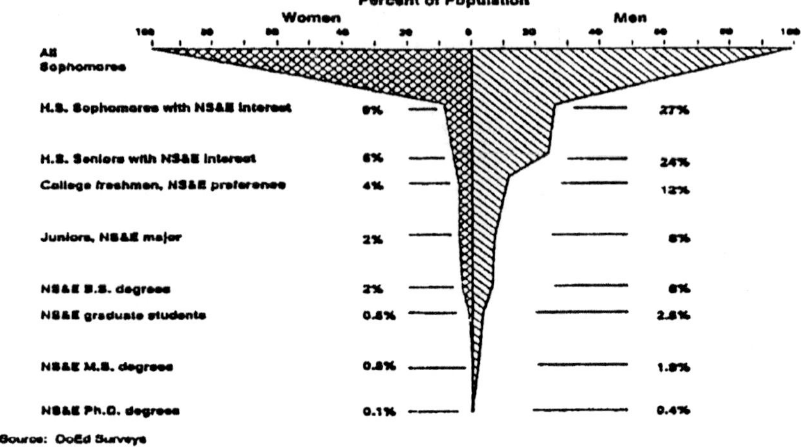


Participation in Natural Sciences & Engineering by Ethnic Group

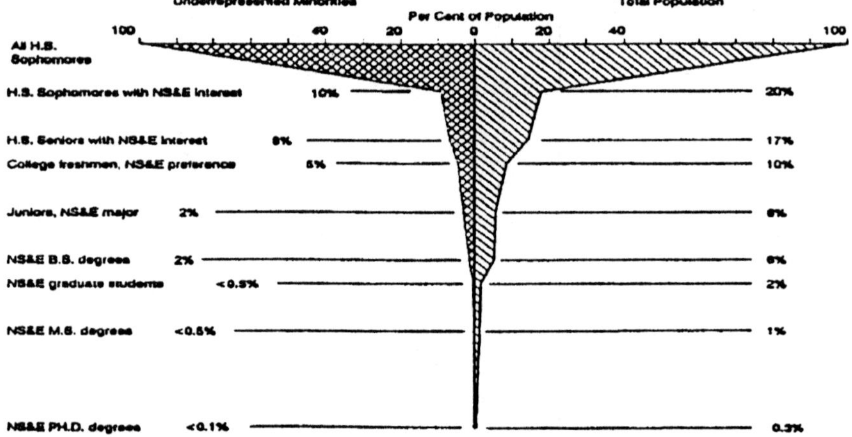


**Figure 4.4. Pipelines for Women and Minorities in Science and Engineering.**
Source: National Science Foundation 1988c, pp. 17–8.

government "provide stable and substantial support for effective intervention programs that graduate quality scientists and engineers who are members of underrepresented groups," the Task Force called upon the federal government to become involved in solving the problem of underrepresentation of women, minorities, and disabled persons in science and engineering. By encouraging academia "to set quantitative goals for recruiting and graduating more U.S. students in the sciences and engineering from underrepresented groups . . .[and] to improve retention programs aimed at underrepresented groups . . ." (Task Force on Women 1989, p. 11), the Task Force also called upon academia to vigorously take up the task of increasing the representation of women, minorities, and disabled persons in science and engineering. Educators at all levels began to see themselves, and what they did, as parts of the pipeline (Bailey 1990; Bernes 1992; Anderson 1993; Hawkins 1993). In sum, the Task Force established an exchange of federal monies for numbers of underrepresented science and engineering students. Universities, as they implemented programs to attract and retain women and minorities, now received federal funds.

## The Pipeline as the "The Sputnik of the Eighties"

In 1983, seeking a national crisis that would focus government support into education, Charles Smith, Chancellor of the University of Tennessee, asked in the popular media "Sputnik II—Where Are You When We Need You?" (Smith 1983). In 1987, the NSF's pipeline became the Sputnik of the 80s. According to the Executive Director of the Task Force and Deputy Division Director in Engineering Education at the NSF, after the Task Force final report was presented to Congress "the budgets for education in science and engineering went up tremendously, at least a three or four fold increase." As we have seen, the NSF's funding for education and human resources increased from $110 million in 1987 to $600 million in 1996 (see figure 4.5).

The impact of the Task Force report was so significant that Richard Ellis, Director of the Engineering Manpower Commission of the AAES, testifying during congressional hearings on oversight of the pipeline numbers said: "The Task Force used the scarcity arguments to dramatize its push to open up technical professions to underrepresented people. The *Chronicle of Higher Education* then proceeded to focus as much on the scarcity aspects of this story as on the need to encourage women and minorities. From there, NSF's numbers went out to thousands of college public relations people and then to the media" (U.S. House Committee on Science 1993, p. 246).[22]

By the end of 1980s, the NSF had redefined its mission around economic competitiveness and the pipeline. In its *Long Range Plan FY 1989–1993*, the NSF outlined its contribution to the nation's economic competitiveness and security by supporting activities in three categories: education, the generation of new knowledge, and the transfer of new knowledge from producers to users. Among the three themes to develop this plan, the purpose of one was to "enhanc[e] the quality of science and engineering education and of human resources development, and broaden participation in science and engineering." The other two are improving academic science and engineering facilities, and developing science and technology centers and group research activities. The importance

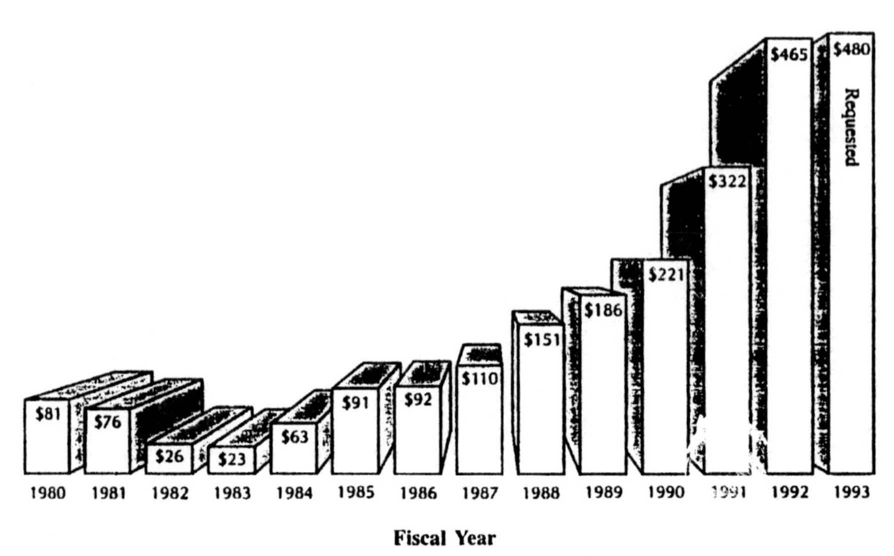


**Figure 4.5. Funding for NSF Education and Human Resources: 1980–93.**
Source: National Science Foundation 1992, p. 5.

now given to developing educational and human resources, particularly those of women and minorities, was clear: "Among the key NSF themes and strategies, this will be the top priority"(National Science Foundation 1988b, p. 9). The report clearly states that the rate of growth in support of education and human resources will be maintained at a level at least 50% higher than that in other strategic areas (disciplinary research and facilities, centers and groups).

The distribution of the budget, and the funding of continuing and new programs was conceptualized along the pipeline. Specific locations for funding along the pipeline were established: elementary/secondary level; undergraduate level; graduate level; postdoctoral level; young investigators; and senior scientists and engineers. Special provisions to increase the participation of women and minorities in each one of these levels were also established.[23] In sum, the pipeline had become good business for the NSF. It helped the NSF become the nation's premier agency in the making of scientists and engineers by significantly increasing its budgets and legitimizing its position within academia and with respect to underrepresented groups.

## WOMEN AND MINORITIES INVOKED A NEW IMAGE OF NATION

Prior to the Task Force recommendations, calls for increasing women and minorities in science and engineering were for the most part unsuccessful. After women and minorities had made some strides in the 1970s, advocates for social justice and equal opportunity continued to fight the Reagan administration's decision to cut science education programs. However, in the early 1980s, they were invoking an outdated image of nation. For example, while trying to save science education programs at the NSF in 1982, Charles Meredith, Chancellor of Atlanta University Center, claimed that "it is extremely significant to realize that the production of minority professionals [in s/e] is essential to the entire framework of achieving both social and educational equity [for minorities]" (U.S. House Committee on Science and Technology 1982b, p. 481). His attempts failed. Other failed attempts include those of John Slaughter, the first black to become NSF Director, and Dr. Sheila Pfafflin, President of the Association for Women in Science Educational Foundation and Chair of the Subcommittee on Women of the NSF's Committee on Equal Opportunity in Science and Technology. In 1982, she had tried to further develop programs for minorities in science at the NSF based on claims of social justice. The point here is that these advocates had been using arguments from the 1970s when Congress had given access to women and minorities to science and engineering as a means of solving domestic national problems, including social and educational equity. However, in the early 1980s, the image of nation had shifted and the nation's needs were being redefined in terms of economic competitiveness. Any group hoping to further its participation in science and engineering would have to speak to these new national concerns. Calls for education programs, even in science and technology, had to be redefined in terms of economic competitiveness.

No one understood this better than Shirley Malcom, who had worked her way up from a Research Assistant position at the AAAS in 1975 to become Director of the AAAS Office of Education and Human Resources and a member of the NSB. Malcom, who in the 1970s had argued successfully that "diversity of cultural and other backgrounds was good for science" (Malcom, Hall et al. 1976), understood that the legislative gains of the 1970s would not ensure the representation of women and minorities in science and engineering education in the 1980s. With the Administration committed to fiscal reorganization and the nation focused on economic competitiveness, Malcom shifted her strategy: to build a need for women and minorities into the discourse of competitiveness. This shift in strategy came very early in the Reagan era when Malcom wrote in an editorial in *Science* magazine: "We must protest cuts in programs for developing the capabilities of women and minorities, not only for the sake of these groups, but also for the sake of science and for the sake of our nation" (Malcom 1981).

Malcom worked to link the issue of underrepresentation of women and minorities to the problem of economic competitiveness in a way that would resonate with everyone, regardless of their political affiliation. U.S. Representative Margaret Heckler (R-Mass), a member of the House Subcommittee on Science, Research and Technology, also realized the need to link underrepresentation to the issue of manpower for competitiveness within the NSF:

> I feel we are frightfully behind. We had to fight very hard to save women and minorities last year. Now we know that on the one hand we have the technology problems, the personnel problems, the academic training needs, the productivity lag between the U.S. and Japan, all these enormous difficulties facing the industry and the jobs affected by it, and here we have an enormous resource in the population of women and minorities and we do not really seem to be making the right linkages. I would hope that there would be a development of a sense of urgency within the NSF on this subject, not only in terms of programs, but on the whole picture. I think we are dealing with fringe issues and not centrally attacking what could be a major resolution of national problems (U.S. House Committee on Science and Technology 1982b, p. 534).

In the beginning of the 1980s, neither Malcom nor Heckler had at their disposal the official knowledge that was to come later with the pipeline. The Task Force did have this knowledge at its disposal and used it to recommend policy and programs.

### Women and Minorities to Save the Nation

During invited lectures, Erich Bloch positioned the NSF at the center of the pipeline industry. Always opening his lectures with the pipeline argument and outlining the NSF's programmatic solutions, he invited academia to participate and created partnerships with industry, all in the name of economic competitiveness:

> NSF has set in place a number of programs to meet the challenge . . . but these programs are only a place to start. Their success depend on you . . . I encourage you to become involved. Reach out to students in your local schools . . . and encourage them to take the prerequisite math and science courses . . . Talk to your colleagues in industry and state and local government about the problems and ask them to help you improve engineering training and to recruit and retain women and minorities in engineering . . .Your success will help determine whether the U.S meets the challenges of a new world environment (Bloch 1989).

During his speeches to audiences comprised of underrepresented groups in science and engineering, Bloch showed the growth of NSF programs for women and minorities after 1985, the year in which he began using the pipeline argument in Congress to secure appropriations for such programs.

By 1988, the National Science Board (NSB) imparted a mission to the NSF to bring together the pipeline, economic competitiveness, and the importance of underrepresented groups. In determining the NSF's role in economic competitiveness, the NSB, in a report titled *The Role of the National Science Foundation in Economic Competitiveness*, stated the following:

> If compelled to single out one determinant of US competitiveness in the era of the global, technology-based economy, we would have to choose education, for in the end people are the ultimate asset in global competition . . . But an especially important further step will be to extend the pool from which the pipeline draws by bringing into it more women, more racial minorities, and more of those who have not participated because of economic, social, and educational disadvantage . . . Thus not only is providing a better grounding in math and science for all citizens a matter of making good on the American promise of equal opportunity. It is a pragmatic necessity if we are to maintain our economic competitiveness (National Science Board 1988, p. ii).

With that in mind, one of the NSB's final recommendations to Congress was that

> From the perspective of economic competitiveness . . . NSF programs and management efforts designed to help bring women, minorities, and the economically, socially, and educationally disadvantaged into the mainstream of science and engineering deserve continued focus (Ibid., p. iii).

In 1984, before the pipeline studies created women and minorities as significant categories for economic competitiveness, the only existing program in the NSF's science education budget for underrepresented minorities was the Minority Graduate Fellowships. At that time, only 10% of the sixty total NSF fellowships were awarded to minorities and women. By 1989, "education programs . . .to strengthen the quality, diversity, and number of U.S. scientists and engineers" became the special focus of the NSF's mission (National Science Foundation 1989). Unprecedented in the NSF's history were the kinds and number of programs specifically targeted to minority populations.[24]

## CONCLUSIONS

In the 1980s, an image of nation under the threat of Japanese competitiveness emerged. Pro-technology groups, including industry leaders and "technophiles" such as Lewis Branscomb, rushed to define this new national problem and its solutions on their own terms, setting the limits of discourse. They defined the problem as that of failing productivity, and the solution in terms developing technological innovation and the education and training of human resources, mainly engineers. Both the congressional attempts to advance a national manpower policy and Reaganomics' redefinition of the government's role in education defined the limits of what was sayable about education and the nation. Within these limits, the NSF emerged as an institutional solution to the political problem of educating scientists and engineers for economic competitiveness.

Erich Bloch replaced the last of the scientific academists and directed the NSF's transformation in order to meet the needs of economic competitiveness by developing technological innovation, infrastructure, and workforce. To conceptualize the nation's new needs for scientists and engineers, he implemented and successfully institutionalized an engineering flow model: the science and engineering pipeline. More than simply becoming the guide of the NSF's education programs, the pipeline became the "Sputnik of the Eighties." First, it gave a visual form to the problems of education (such as quantified "leaks" from kindergarten to Ph.D.). Second, it showed the potential contributions of large numbers of scientists and engineers (quantified future production of scientists and engineers) to economic competitiveness. The NSB, Congress, the President, and educators embraced the message of the pipeline and established a new social contract between the federal government and those groups that were needed to feed the pipeline, mainly women and minorities. This new contract created what I have called the pipeline industry, in which pipeline-based claims allowed educators and NSF administrators to request federal monies to bring students into the pipeline by means of recruitment and retention efforts. Women and minorities in science and engineering, after disappearing in the early 1980s, re-emerged in the late 1980s as significant categories for economic competitiveness.

## NOTES

1. Early examples of proposed legislation for national policy include "H.R. 7130: National Engineering and Science Manpower Act of 1982" and "H.R. 1310: Emergency Mathematics and Science Education Act of 1983." It is significant that the word "engineering" is placed first in the title of H.R. 7130, which was proposed to establish a national policy to be administered by the NSF. The word "engineering" is missing from the title of H.R. 1310 which was proposed to be administered by DoE.

2. One of these proposals called for a Coordinating Council on Engineering and Scientific Manpower, which included the Secretary of Education, the Director of NSF, the Director of OSTP, the Director of OTA, and other members appointed by Congress and the President (U.S. House Committee on Education and Labor 1982).

3. The NEAC's participants included: the American Council on Education, the Association of American Universities, the Association of Independent Engineering

Colleges, the National Association of State Universities and Land-Grant Colleges, the Business-Higher Education Forum, the Business Roundtable, the Committee for Economic Development, the Council on Foundations, the National Association of Manufacturers, The Conference Board, the ABET, the AAAS, the American Association of Engineering Societies, the American Electronics Association, the American Society for Engineering Education, the Industrial Research Institute, the National Academy of Engineering, the American Association of Small Research Companies, Licensing Executives Society, National Action Council for Minorities in Engineering, the Committee on Science and Technology of the U.S. House of Representatives, the Office of Science and Technology of the Executive Office of the President, NASA, the NSF, the U.S. Department of Defense, and the U.S. Department of Energy.

4. Founded in 1978, in affiliation with the American Council on Education, the Forum is an organization of leading corporate and academic chief executives. Among these we have CEOs from companies like Westinghouse, General Electric, General Motors and Presidents from universities such as Harvard, John Hopkins, Duke, U.C.L.A., and Cal Tech.

5. The reaction of academic administrators and educators, and of government and industry leaders, is reflected in an avalanche of articles that followed *A Nation at Risk*. See for example Hogg 1983; Howe 1983.

6. For FY 1983, Fellowships received $15 million, while pre-college education received $1 million. Knapp directed no money into undergraduate education.

7. Among the members of the Committee, there were deans of engineering from prestigious U.S. universities, vice presidents of Engineering from large corporations, and Executive Directors of engineering from professional societies. There were also numerous engineering faculty, too many to list here. For a complete list, see National Research Council 1985.

8. The Commission members included powerful actors from industry, government and academia such as: John A. Young (Commission's Chairman), President and CEO of Hewlett-Packard Co., Robert Anderson, CEO of Rockwell International, George Keyworth, Science Advisor to the President, George Low, President of Rensselaer Polytechnic Institute, Michael Porter, School of Business Administration at Harvard University, Mark Shepherd, CEO of Texas Instruments, Inc., and Egils Milbergs (Commission's Executive Director), Deputy Assistant Secretary of the U.S. Department of Commerce.

9. In 1984, President Reagan nominated Erich Bloch as NSF Director and Nam Suh as Assistant Director for Engineering (former Director of Laboratory for Manufacturing and Productivity at MIT). Suh took NSF's engineering into new directions following NAE recommendations about new divisions based on critical technologies. See National Science Foundation 1985.

10. The PRA existed at NSF since 1974–5, when NSF Director Guyford Stever became Science Advisor to the President after the abolition of the OSTP by President Nixon. In line with the 1970's national needs, the PRA focused on technology transfer policy and energy R&D policy. In the mid-80s, with Bloch as Director, the focus was on policy research that reflected an emphasis on free-market and economic competitiveness, such as the impact of tax laws on industrial R&D, or government support of high technology in several industrial nations (National Science Foundation 1984).

11. It is important to note that career choices in computer-related fields were left out of the assumptions. Enrollment in those fields was at an all-time high, and taking this choice into consideration would not have resulted in a manpower crisis.

12. For example, the PRA stated that "new females baccalaureates in NS&E fields grew steadily from 1.5 % of the female 22–year-olds to 2.4 % during the 1970's. In the first five years, the growth was entirely in life science, while in the latter five years, growth was entirely in the remaining NS&E fields. During 1982–86, female NS&E baccalaureates rose to 3.2 of female 22–year-olds, with the growth almost entirely in computer science" (National Science Foundation 1990, p. 191).

13. The PRA made similar assumptions for the supply and demand of Ph.Ds. For example, it claimed that, for all U.S. citizens receiving baccalaureates in science and engineering, the continuation rate from B.S into Ph.D. was fixed at 5%.On the demand side for Ph.Ds, the PRA claimed that "demand will increase due to new research positions in industry and academia mainly driven by massive retirement of faculty from the boom period of the 1960s."

14. In the early years of pipeline development, the lack of consensus within the NSF was exemplified in the PRA's inability in 1987 to secure an NSF number for its first official report. Although the report, *The Science and Engineering Pipeline*, was published with a PRA number, the *1987 Science and Engineering Indicators*, the NSF's most visible and most well-known official statement about the state of science and engineering, contained a section on the pipeline. By 1989, *The State of Academic Science and Engineering* was the first official NSF publication (with an NSF number) containing the complete argument of the pipeline studies and was published by direct request of Erich Bloch.

15. Throughout this time there was disagreement over demographic-based modeling. The SRS published a report, *Projected Response of s/e/t labor market to defense and non-defense needs* and the OTA published *Demographic Trends and the S/E Workforce*. Both underplayed the importance of demographics in the shortages of scientists and engineers.

16. See House and Shull 1991 where they justify the use of PC technology for policy analysis under the argument of democratizing the policy process.

17. See, for example, the PRA Issue Paper (84–53) Demand-Supply Balance for Scientists and Engineers Within Government, Industry, and Academic Sectors. Also PRA. 1988. Personnel in Natural Science and Engineering: Working Draft prepared for NAS Government-University-Industry Research Roundtable (GUIRR) Working Group on the Research University Enterprise.

18. As Assistant Program Manager in Engineering Education, while preparing speeches and presentations that promoted NSF engineering education programs to different audiences, I also prepared a variety of future scenarios for different engineering disciplines. To produce a custom-made pipeline, one only needed to use discipline-specific data coming from the SRS, the NSF's statistical unit, and plug it into the programmed pipeline model.

19. Around this time, many other task forces, specific to certain population groups in specific disciplines, such as "Women in Engineering," were being formed. How-

ever, the significance of the one mentioned here is that it was the only one given authority by Congress and established under public law (99-383).

20. Task Force Co-Chairs: Jaime Oaxaca, Vice Chairman, Coronado Communications, and W. Ann Reynolds, Chancellor, California State University System. Executive Director (appointed by Erich Bloch): Sue Kemnitzer who eventually became Deputy Division Director for Engineering Education at NSF.

21. According to the Task Force, those sectors are the President, industry, state legislators, school boards, parents, the media, governors, the federal government, universities and colleges, pre K-12 educators, professional societies, and all Americans.

22. An abundance of reports was published after the pipeline's introduction, using its claim of shortage and creating a national manpower crisis. See for example, Aerospace Education Foundation 1989; Holden 1989; Atkinson 1990; Shaw 1990; Vaughn 1990.

23. Examples of these programs include Visiting Professorships and Research Opportunities for Women, Minority Research Initiation Grants and Minority Research Centers of Excellence, Minority Graduate Fellowships, and Research Assistantships for Minority High School Students.

24. These included pre-college activities such as "Escalante's Work Expands" and the Detroit Area Pre-College Engineering Program, and college-level activities such as Undergraduate Research Opportunities for Minorities, the Minority Scholars Program, Women in Engineering Fellowships.

## *Chapter Five*

# America in the "New World Order": Making Flexible Scientists and Engineers for Global Competition

> . . . the time has come, for the first time in U.S. history, to establish clear, national *performance* [educational] *goals* that will make us internationally competitive. . . . By the year 2000, U.S. students will be first in the world in mathematics and science achievement . . . (National Science Foundation 1991, p. vi) (italics mine).

> . . . we need human resources that will be *flexible* enough in terms of their training so that if they don't quite match what is at that time the need for their skills, they can be retooled very quickly (Fechter quoted in (U.S. House Committee on Science and Technology 1985, pp. 64–5) (italics mine).

In the 1990s a new image of America emerged. This shift was due in part to new understandings of Japan and other Asian nations not in terms of numbers but in terms of characteristics of their cultures and government-industry relations that allow them to innovate and compete. At the same time, unified Europe emerged as a political and economic reality that threatened the economic supremacy of the US. The new image challenged American institutions and policymakers to reconceptualize US scientific and engineering institutions, policies, and programs conceived during the Cold War to match the requirements of new global competition. NSF's programs for research and education were certainly no exception.

The apparent contradiction between the first quote's call for national performance *standards* and the second quote's call for *flexibility* in training of human resources reflects how the new image came to challenge education policymakers. Both standards and flexibility became the dominant discourses to educate and train U.S. scientists and engineers for global competition. This chapter traces the emergence of a new dominant image of America challenged by global competition and its influence on the NSF, its policies, and programs in education and human resources. During the 1990s, the NSF and its programs shifted from producing large numbers of scientists and engineers to beat the Japanese towards making flexible scientists and engineers capable of rapidly adjusting to changes in knowledge production, dissemination, and application in new global scenarios. Focusing on two emerging discourses for education and human resources in science and engineering—systemic reform and flexibility—, this chapter shows champions of these two discourses defined the limits of what was possible for policies and programs to educate and train a highly skilled workforce to help the U.S. compete in new global scenarios. Furthermore, this chapter analyzes how these new policies and programs threatened to redistribute power in U.S. science and engineering education, from groups based on gender and race to groups based on performance, and the pervasiveness of the pipeline metaphor and the end of the 1990s.

## THE EMERGENCE OF A NEW IMAGE: GLOBAL COMPETITIVENESS THREATENS AMERICA

In the 1990s popular and academic media portrayed the global marketplace not as a "tug-of-war" for market-share of consumer products between Japan and the U.S. but in terms of multi-lateral "battlegrounds" where the growing economies of unified Europe and the Pacific Rim became significant U.S. competitors and unexploited markets. Since the late 1980s, both media and academic writers began remapping the world from a place where only two opposing sides define the geopolitical context—first, U.S. vs. USSR, and later U.S. vs. Japan—to a place where three or more "poles" share and compete for global political and economic space. Popular media descriptions included the U.S. competing with the rest of the world, the U.S. competing with emerging small but powerful economies of South East Asia, and the U.S. being threatened by European unification (1991; 1992; 1992). Academic descriptions included a world made of three economic centers with Europe as the "next battleground" where Japan and the U.S. will have to compete for market share, and a global space continuously changing according to ongoing redefinitions of competing or collaborating alliances between U.S., Europe and Japan.[1] Among these descriptions, two locations clearly emerged as the new economic threats to the American nation: the Pacific Rim and a united Europe.

Along with geopolitical and economic concerns came worries over how cultural differences translated in the capacity to innovate and be competitive.

Descriptions of the Pacific Rim showed a block of emerging economies—often called "little dragons" or "seven tigers"—as smaller versions of Japan, which through a combination of industrial policies, cultural values, and human resources became serious competitors to the US.[2] Although reports on Japan and other Asian countries showed significant differences in the economic policies of these nations, when trying to come up with answers to these countries' economic and technological successes, most writers provided one answer: Confucian culture. As one leading engineering magazine put it: ". . . the Confucian ethic of toil, respect for authority, and preference of cooperation over confrontation, has inspired [in these countries] achievements far out of proportion to those of [Western] nations" (Rosenblatt 1991, p. 25).

In spite of the recognition that cultural differences might be the key to the emerging competitive advantage of Asian nations, concerns over differences in the number of scientists and engineers did not go away. However, when the shortages of scientists and engineers predicted by the pipeline of the 1980s did not materialize and the emerging image turned people's attention to culture, calls to increase the number of scientists and engineers became questionable. For example, Earl Kinmonth questioned the assumptions of most number-based studies. Kinmonth claims that "unfortunately, the notion of Japanese numerical superiority is entirely a myth. Every component of it can be shown to be the product of differences in definition and methodology between U.S. and Japanese data sources. Nevertheless, because Japanese engineering success remains a fact, attention must shift to qualitative differences in engineers [i.e., education and culture] and how they are used in industry [i.e., management]" (Kinmonth 1991, p. 329).

Along with the questioning of quantitative comparisons came an obsession to explain Asian success in terms of Confucian culture and industrial policy. As Gail Cooper explains, "In the now bulky literature that seeks to explain Japan's postwar 'economic miracle,' two broad explanations have emerged. One is that Japan's success is due to its cultural base . . . recently dubbed 'Confucian Capitalism', an explanatory model relevant for all Asian industrialization. . . . A second group maintains that cultural explanations prevent us from seeing that specific modern policies rather than a centuries-old culture nurtured Japan's economic growth" (Cooper 1993, p. 207). As we will see, US educational policymakers came to respond to these concerns over the cultural and policy dimensions of the new threat through systemic reform and flexibility programs.

Along with describing the emergence of Asian powers in cultural terms, both popular and academic writers redefined a unified Europe as a multinational/cultural space coming together and becoming not only a competitor of the U.S. for world markets but an unexploited market that both Japan and the U.S. will have to fight for. With Europe rapidly becoming a single economic market and "the world's largest market and largest trader" (Bergsten 1990; Jackson 1993), a view of the world defined around two rival sides was no longer valid. Bergsten gave this new global arrangement the name of "collaborative tripolarity" where "the Big Three of economies will supplant the Big Two of nuclear weaponry on the issues that will shape much of the early twenty-first century. Japan, a united Europe, and the United States will become full partners in managing the world economy" (Bergsten 1990, p. 653).

The challenge of the new image of America under the threat of global competition led some political theorists to analyze state-societal arrangements, particularly Japanese and German ones, in order to explain their recent successes in international competitiveness. For example, according to Jeffrey Hart "the Japanese system is well organized for joint state and business efforts to bring Japan to the technological frontier in strategic industries and keep it there" (Hart 1992, p. 284). To account for Germany's success, Hart argued that "the strength of the German system is built on the high skill level of German workers. . . . Although the German government plays a minor role relative to others, it is responsible for the educational system that transmits skills to the workforce and helps to assure the transmission of university-created knowledge to business." Hart concluded that "state-societal arrangements are the key to explaining recent changes in international competitiveness" (Ibid.). This concern for global competitiveness in terms of state-societal arrangements led political theorists and U.S. policymakers to ponder what kind of new arrangements should be developed between the federal government and society in order to enhance US competitiveness. In Hart's own words, "If [the U.S.] chooses the Japanese model there will have to be a major upgrading of government agencies and centralization of industrial policy-making in a single agency. . . . If the U.S. chooses the German model . . . a significantly increased commitment to the training and retraining of workers will be needed" (Ibid., p. 291). As we will see, whether by design or not, US educational policymakers ended up emulating the German model by creating policies and programs aimed at facilitating skill transfer to science and engineering students, and knowledge transmission from university to business.

These new calls for the U.S. government to redefine its state-societal arrangements had an impact on the position of the NSF within the federal government. We have seen how since the 1980s different pro-technology groups from industry, academia, and government tried unsuccessfully to make the NSF an American version of Japan's MITI. Although NSF did not, and probably never will, become an agency that dictates centralized industrial policy, it became instrumental in the 1990s in orchestrating partnerships between government, industry, and academia in both research and education.

Also, in German-like fashion, the NSF created programs to educate and train the American labor force into a highly skilled workforce while ensuring the transmission of university knowledge to business. In techno scientific education, NSF tried to address both the policy and cultural dimensions of the Asian and European educational challenges by means of two major programmatic initiatives: systemic reform and flexibility.

## SCIENCE AND TECHNOLOGY FOR GLOBAL COMPETITION

On March of 1993, the Subcommittee on Science of the House Committee on Science, Space, and Technology held congressional hearings under the title "The Mission of the National Science Foundation." The Subcommittee's chairman, Rick Boucher (D-Va) sought to redirect NSF's mission towards new national needs for global competition. Boucher recognized that "as new opportunities and challenges have been created by the end of the Cold War, the rise of multilateral economic competition from abroad, and the emergence of global environmental problems," NSF's mission needed re-evaluation (U.S. House Committee on Science 1993a, pp. 1–2). The main guide for NSF's new mission was the testimonies of three powerful groups who were being challenged by the new image of the American nation: the President Council of Advisors on Science and Technology (PCAST), the Carnegie Commission Task Force, and the Government-University-Industry Research Roundtable's (GUIRR) Working Group of the National Academy of Sciences. With significant membership from the technophiles of the 1980s, these groups published reports on the future of U.S. science and technology in the new era of multilateral competitiveness and their recommendations significantly influenced the future mission of NSF.[3] Redefining the limits of what was sayable about science and engineering in relation with global competitiveness, the three groups claimed that techno-scientific knowledge production, and the development of scientists and engineers to produce it, were fundamental for ensuring the competitive position of the U.S. in the new global scenario. As the PCAST report stated: "advancing the frontiers of knowledge is not, as it once may have been, a matter of intellectual luxury. *In an era of relentless global economic competition, it is a national imperative*" (U.S. House Committee on Science 1993a, p. 90) (italics mine).

The new limits of discourse also called for the production of techno scientific knowledge to take place in multinational settings beyond the walls of U.S. research labs and national borders, and to be produced by multinational research teams. For example, the GUIRR Working Group, with former NSF Director Erich Bloch as its chairman, proposed a "Global Research System" in its report *Vision for the Future*. GUIRR argued that as "international research cooperation will become a pervasive feature of the U.S. academic research enterprise in the next century, multinational research arrangements will be essential for studying large-scale [scientific and technological] problems. The research communities of both industrialized and developing countries will rely more and more on cooperative ventures to address these and other research problems" (Ibid., p. 257). It also became clear that scientific and technological cooperation was just another means towards strengthening the U.S. competitive position. In justifying multi-national cooperation, the GUIRR report stated:

> Global competition in science and technology will require that the United States pay close attention to the research activities of other countries, especially those targeting economic growth as their primary research goal. This will be particularly true for the Western European and Pacific Rim countries, which have become fierce competitors in the knowledge-intensive global marketplace. . . . Just as Japan in past decades capitalized on discoveries made in this country, during the next century, U.S. universities and industries will benefit from the growing base of knowledge and technology produced elsewhere (Ibid., p. 258).

Following this kind of recommendations NSF acknowledged the need to establish international partnerships. In its 1995 report *NSF in a Changing World*, NSF recognized that "our goal of world leadership requires that we carry our partnerships across national boundaries, working with comparable organizations in other countries to promote international cooperation wherever mutually beneficial." To promote this sort of competitive cooperation, NSF proposed two kinds of funding: institutional and individual. Through institutional support, NSF sought "[to] enable and encourage U.S. scientists, engineers, and their institutions to avail themselves of opportunities to enhance their research and education programs through international collaboration." By means of individual support, NSF aimed at "provid[ing] future generations of U.S. scientists and engineers with the experience and outlook they will need to function productively in an international research and education environment through support for traveling fellowships and research activities at overseas sites" (National Science Foundation 1995a, pp. 19–22).

As response to the European and Asian challenges, NSF opened offices in Tokyo and Paris "to represent NSF in the science organizations in Europe and Asia, to monitor and report on those developments in Europe and Asia that can impact American science and engineering [including education and human resource developments], and to identify and develop new opportunities for collaboration between the U.S. and European science and engineering communities" (National Science Foundation 2004).

## Human Resources

In their analyses of U.S. competition with Europe and the Pacific Rim, scholars of competitiveness identified education and human resources as significant elements of global competition. They recommend investment in education and human resources as safe policy that enjoyed political support at home and that was non-confrontational, as some industrial and trade policies are. For example, recommending education and training to any of today's global competitors, Tim Jackson claimed that "whether in Washington, in Brussels, or in the different national capitals of the European community . . . governments [should] foster good human capital, by turning out a highly educated and skilled worked force from national schools" (Jackson 1993, p. xiv).

As a key to economic competitiveness, education became reconceptualized not in terms of numbers but in terms of cultural dimensions after more than half a decade of comparing US education with Japan's. According to educational experts, Japanese family-rooted values—such as high regard for the group, hard work, diligence, and perseverance—are reinforced through education and translate into educational performance, loyalty to the company, team work, and quality. In short, education reinforces the values that are key to Japan's economic competitiveness. As a U.S. Department of Education report titled *Japanese Education Today* stated, "Japanese education is a powerful instrument of cultural continuity and national policy. The explicit and implicit content of the school curriculum and the manner in which teaching and learning are accomplished impart the attitudes, knowledge, sensitivities, and skills expected of emerging citizens of Japanese society. These lessons are further reinforced in the context of family and society" (U.S. Department of Education 1987, p. 2). As former Secretary of Education, William J. Bennett asked, "What lessons might we draw for ourselves from a close look at Japanese education? It is scarcely a novel query. Japan, after all, has increasingly become a reference point or gauge by which Americans appraise our own education system" (Ibid., p. 69). By contrasting Japanese collectivism with American individualism, educators began to incorporate collective traits, such as teamwork, in American curricula. Other numerical and cultural comparisons between the U.S. and Japan led influential business and academic leaders to conclude that the key to Japan's productive performance resides in the quality and flexibility of its workforce, and that these traits must be transferred to U.S. education and training of scientists and engineers.[4] According to these leaders, the new challenge to education and training of human resources was to find ways to transfer and institutionalize those "cultural tricks" in order to improve the performance of new capitalist modes of production while preserving individualism to maintain an ever-growing consumer market(Morishima 1984; Feinberg 1993; Lorriman and Kenjo 1994). NSF responded to these calls through programmatic initiatives aimed at developing human capital: *systemic reform* and *flexibility*.

## SYSTEMIC REFORM

Born out of the *Goals 2000: Educate America Act of 1993*, systemic reform, according to NSF, is a set of "fundamental, comprehensive and coordinated changes made in science, mathematics and technology education through attendant changes in school policy, financing, governance, management, content and conduct" (National Science Foundation 1995b). These changes were to be achieved through three interconnected aspects of systemic reform: 1) unifying *visions and goals*, including high standards for learning expected from all students; 2) *alignment among all parts of the system*, including policies, practices and accountability mechanisms; 3) a *restructured system of governance and resource allocation* that places greatest authority and discretion for instructional decisions on school sites (National Science Foundation 1996).[5]

### The End of the Pipeline?

On April 1992, the Subcommittee on Investigations and Oversight of the Committee on Science, Space, and Technology, U.S. House of Representatives, held congressional hearings to determine "How Good are the Numbers?" of the projections made by NSF's Division of Policy Research and Analysis (PRA) in their pipeline studies(U.S. House Committee on Science 1993b). House Representative Howard Wolpe (D-Mich), chairman of the subcommittee, stated that "the purpose of the hearings is to review how a study so flawed survived for so long in the nation's premier scientific agency. The subcommittee's investigation has revealed that valid criticism was ignored and even suppressed within the Foundation" (Ibid., p. 2). Wolpe was referring to PRA's pipeline studies and their claim of a shortfall of scientists and engineers that would put the U.S. behind in the competitiveness race.

Wolpe and many critics claimed that the media, both popular and academic, took PRA's projected shortfall of 675,000 scientists and engineers by the year 2006 and made it into a crisis of national proportions. However, after dozens of witnesses from both sides of the issue presented their testimonies, and even the recommendations from an impartial expert were brought in, this controversy had no major impact on NSF's credibility or in its budgets appropriations. In the end, all NSF had to do was to correct the record stating that shortages are not the same as shortfalls and demographic projections are not the same as market analyses. This controversy over the pipeline did not have major consequences for education and human resource funding because a different image of nation had already emerged, one that called for flexibility of the workforce instead of for large numbers. The studies were dismissed but the metaphor lived on.[6]

Around 1993–94, advocates of both systemic reform and competitive flexibility had already made their entrance in NSF. Challenged by the new image of the American nation, these new visionaries, and even some important advocates of the pipeline in the 1980s, were beginning to criticize the pipeline model for promoting specialization at times when flexibility of the workforce was paramount. The first criticism came from advocates of systemic reform who proposed to engage and reform all parts of the educational system simultaneously. Systemic reform advocates criticized the pipeline for its narrow focus on those who stay in, and how to keep them in, while ignoring those who stay out of science and engineering education, regarding them as "leaks" or "drop-outs." These limitations were recognized by a former director of NSF's Division of Research, Evaluation, and Dissemination (RED), who tried to change the ways in which NSF viewed its educational goals. Explaining the limitations of the pipeline, this former director said

> the pipeline metaphor is not a very useful metaphor . . . because it restricts at various entry points and there aren't too many of them. There are still more people flowing out than people flowing in. It restricts the pool of people who could eventually go on and take degrees in science and engineering. So if you fix the pipeline, if you were to seal the leaks, you are still dealing with a very small minority of all the students and that's the reason that systemic reform is a different kind of approach (interview with author).

Also aware of these limitations were people who made their careers at NSF constructing official knowledge and implementing pipeline programs. For example, a high-ranking official involved in NSF's engineering education programs argued that the pipeline ignores non-linear paths into science and engineering education:

> there is one big flaw of the pipeline. . . . There are a lot of people who do not go on a linear path through school, especially after grade 10. I know an American Indian woman who drops out of school in 10th grade and has finally come back at age 35 and gets a GED and goes to the community college and at age 40 has an associate degree and by age 43 has a bachelor's degree . . . we have more and more of those kind of people. . . . If you look at the average student in the California State University System or the Texas A&M system, they are not 18-year-old white men. The average profile is that of a woman who is 25–35 and that doesn't jive with the pipeline analogy. That is a big problem for science (interview with author).

People in non-linear paths were ignored under the assumptions of the pipeline of the 1980s which focused on the 18–22 year old college population. The goal then was to enlarge the fraction of this population group going into science and engineering. In contrast, systemic reform proposed a more inclusive approach to education for it takes into account all kinds of students that NSF's programs have traditionally neglected. Arguing for the inclusiveness of systemic reform while criticizing the limitations of the pipeline, the former director of RED said that

> the pipeline became a problem of how do you retain identified and self-selected students at the undergraduate level so they can complete a degree and maybe go on to graduate school. In some ways the pipeline is a far easier problem than the one that EHR is addressing. Our commitment is to all students. We are looking at the 95% that NSF traditionally has not worried about. We are now trying to address the needs of all students regardless of what they are going to do both educationally and occupationally (interview with author).

But more than a concern with elitism, these critics were also concerned with the overspecialization of students produced by the pipeline. As the same director of RED pointed out, another problem with the pipeline is that

> it only projected one model of what to do with a Ph.D. in science, namely research. . . . It is our responsibility to diversify what a doctoral program does. Just the way we have been talking about entering the workforce at earlier points, when you end up with a Ph.D. you should be versatile not over specialized. You should be able to go into industry . . . into administration . . . into research . . . into teaching. . . . And the problem here [at NSF] is that we haven't been very creative about preparing students for a range of possible futures after they have a Ph.D. . . . Even if you can get the Ph.D. awarded, you still are not ready to do anything that society at the present time knows how to utilize. I find that to be, 30 years later, a supreme irony (interview with author).

Concurring with this view, the high-ranking official in engineering education viewed the product of the pipeline as less important for economic competitiveness than techno scientific literacy of potential high-skilled workers: "Ph.D.'s don't matter much to economic competitiveness . . . I rather have a much higher literacy in technology and science from people that graduate from high school than get 10 more Ph.D.s" (interview with author).

Despite these criticisms there are still many NSF administrators at research directorates and university officials who still the view the pipeline as serving one of NSF's main goals: the production of Ph.D.'s. After all, this has been the most everlasting education function of NSF, exemplified by its graduate fellowship programs which have always remained at NSF even when science education was zeroed out. When asked if, after the demise of the pipeline studies in 1993, there were still people at NSF who believed that the mission of the agency was the production of Ph.D.s for research, the director of RED responded:

> The vast majority. The only way you are going to hear things you are hearing from me [systemic reform] is if you talk to few other in EHR [Education and Human Resources Directorate]. . . . There are even people who are in this directorate who are still firm believers in the pipeline and think that's what we are all about. So metaphors work. It's a grip that is hard to break (interview with author).

John Jordan describes metaphor as a tool that "creates consensus that would be impossible if participants had to reach explicit agreement on definitions" (Jordan 1994). This characteristic of metaphor allowed the pipeline argument to disseminate and achieve high levels of consensus that in order to compete with the Japanese workforce, America needed to recruit and retain large numbers of students, mainly from underrepresented groups in science and engineering. We have seen how the pipeline metaphor helped NSF administrators to disseminate this message and in doing so to further establish the limits of what was sayable about education and human resources in science and engineering. The pipeline metaphor helped them do this without, as Kinmonth later argued, reaching agreements on whether Japanese defined engineers the same way that Americans do. Systemic reform advocates recognized this strength of the pipeline metaphor and understood that they needed a metaphor of their own in order to convey their message and began establishing the limits of what is sayable about education in America in the 1990s. Although crystallized in the mental models that many have about science and engineering education, the pipeline metaphor did not work for systemic reform because it went against their goal of developing life-long skills in human resources in science and technology. In a white paper, proposing a framework for national policy to the NSB, systemic reform advocates acknowledged that

> while we discard the [pipeline] metaphor here, we have nothing yet to replace it. Continuing use of "pipeline," however, is inimical to what a national policy is trying to achieve, namely, an emphasis on how to draw students to SMET as a major or a set of life skills, not how many "leak" out of the pipeline or opt for majors and careers other than SMET (National Science Foundation 1994, p. 3).

Systemic reform advocates presented this paper to the NSB to be adopted as NSF policy but the NSB rejected the proposal. A senior member of the NSB and an advocate of systemic reform provided a clue as to why systemic reform had difficulties reaching the consensus of the NSB. This advocate acknowledged that the drawback of not having a metaphor would prevent the creation of a coherent vision within NSF as to how systemic reform works:

> I don't think that systemic reform has probably gotten over to any of the directorates yet down at the program officer level. I don't think there is a coherent vision of systemic reform even in the places where it happens to exist. I don't think that any of us have a complete picture of what it means. We have an intuitive sense that it's right because of the approach that we used to apply—a little over here, a little over there—didn't work (interview with author).

This meant that systemic reform advocates were unable to set the limits of what was sayable. Coming up with the appropriate metaphor, one aligned with the emerging image of the nation in the 1990s, was an important strategy to set those limits. Since systemic reform is about interrelating all parts of the educational system, "interrelatedness" seems to be an important characteristic for a possible metaphor of systemic reform. For example, while one high-level advocate of systemic reform proposed an "ecosystem", claiming that people move a lot out of science and engineering and should be allowed back in, while another proposed an "aquafier, a system of water underground that flows from various places to other places, and it comes from many places. The point is that the educational system is not the only source of talent." The later also argued for efforts to bring students into science and engineering from the most unexpected places. Hence, he proposed a metaphor that would reflect these efforts: "Another alternative is the cultivation model where you take an entire set of resources, minerals, oil, or whatever and you cultivate. If you don't find resources [where you expected], you look for nuggets in places where you don't expect to find them and you work with them." More recently activists for diversity in science and engineering education have begun to question the usefulness of the pipeline metaphor and to propose alternatives like that of "a river with feeders, streams, tributaries, oxbows, deep pools and shallow riffles, etc." which better reflects the complexity of pathways in and out of engineering education and employment.

In sum, the pipeline studies might have lost their credibility after the 1993 questioning of PRA's numerical projections, but the metaphor still persists as the preferred way to conceptualize, visualize, and even criticize the educational system. Systemic reform advocates continue to struggle to find a metaphor to achieve consensus for their programs and policies. This struggle between the pipeline and systemic reform advocates shaped the limits of what is sayable in America about science and engineering education in the 1990s. However, as we have seen, actors and groups alone do not set those limits. They had to speak to the existing image of the nation. In the 1990s, those who spoke to an image of a nation under global competition in need of flexible human resources were more likely to set the limits. We have seen how GUIRR did that with respect to defining the limits of how research is to be carried out in multinational arenas and ABET would do that for engineering education. Within NSF, systemic reform advocates found the language to invoke this image of nation through flexibility.

## The Unintended Consequences of Systemic Reform

A significant unintended consequence of systemic reform, whether in the form of K–12 standards or ABET accreditation process, was the reconfiguration of categories of potential scientists and engineers, from groups based on gender and race to groups based on performance, potentially disempowering population groups created by the pipeline studies of the 1980s. Aiming at reforming all parts of the educational system simultaneously, systemic reform was a significant departure from the pipeline framework which only considered the limited population group on the path towards scientific and engineering careers. Systemic reform advocates criticized the pipeline for basing policy and practice on the "innate ability" paradigm, hence, "often [making] judgments about children . . . on the basis of *educationally irrelevant criteria, including socio-economic status, ethnic group, and gender*" (National Science Foundation 1994, p. 55) (italics mine). By shifting the focus from these "irrelevant" criteria to "performance" criteria between disadvantaged and advantaged students, irrespective of their gender and ethnic group, systemic reform could take power and resources away from those groups that gained access to science and engineering education as underrepresented groups based on gender and race.

An essential element of systemic reform, techno scientific literacy became one of the measuring sticks of economic competitiveness.[7] Techno scientific literacy was defined under systemic reform as flexibility in using one's skills to apply techno scientific knowledge in different contexts. As the director of RED pointed out:

> Literacy is not a matter of what you know, it's a matter of what kinds of skills you have that would allow you to apply knowledge that we would call math and science knowledge in this context and be able to apply it in another context. So in that sense it does become a work skill, becomes a competency. And we don't measure that very well, by the way, as you know, most of our literacy measures have to do with what is DNA? . . . what we need to be able to measure is how people use the content that comes out of science and mathematics. And do they understand the processes involved and how you get that knowledge, and how you refine it, and advance it (interview with author).

One of the ironies of the pipeline of the 1980s was that although it aimed at promoting economic competitiveness, not equal opportunity, women and minorities groups wanting access to science and engineering reaped most of the benefits of the pipeline industry. As a result, the pipeline of the 1980s became an instrument of power for those groups, such as National Association of Minority Engineering Program Administrators (NAMEPA), that became committed to the diversification of U.S. scientific and engineering institutions by means of recruitment and retention efforts.

This struggle between pipeline and systemic reform advocates within NSF extended beyond NSF into national forums on science and engineering education for underrepresented groups. Tensions emerged between those who made their careers in the pipeline industry and those who advocate systemic reform. Referring to an appearance at the 1994 NAMEPA conference "Partners in the Pipeline" where systemic reform was presented to an audience of administrators of minority programs in U.S. colleges of engineering, a senior member of the NSB described the audience's negative reaction as

> Fear! because they don't know what it means to lose the minority tag . . . they don't think that they would be the first people consulted about how to make things work for everyone. They might be right, but I think that the question they have to ask themselves at some point is: is this about my job? or is this about this effort? . . . these programs [Minority Engineering Programs] are unlikely to remain in this kind of environment . . . I would hope that they basically start to become offices of student support and instructional improvement. . . . They don't understand that their value is even greater [under systemic reform] . . . they are the ones who have a better sense of the problems and possibilities (interview with author).

If achievement in this kind of literacy of 'just-in-time' skill deployment, not gender and race, determined your access to the science and engineering workforce, systemic reform was on its way to disempower those underrepresented groups that the pipeline of the 1980s empowered. If metrics and assessment, as required under systemic reform, determined your level of achievement, these evaluation tools were in the process of shifting power, from recruitment and retention programs to systemic reform experts who set and assess the standards.

## The Pipeline Lives

Advocates of systemic reform never created a technology of government to constitute 'performers' as a significant statistical category in the same way that the pipeline created 'women and minorities in science and engineering.' In accreditation processes that became effective in the late 1990s, systemic reform champions created a process to assess how knowledge delivery materializes in the abilities required of flexible graduates. But these processes do not count nor predict the numbers of scientists and engineers required for global competition. These processes do not create a statistical category or constitute a demographic group. Hence 'performance' lacks demographic support in the political process.

In the late 1990s, powerful supporters of systemic reform such as Luther Williams, EHR Director, left NSF. NSF recently announced that it will change its systemic reform approach to one based on problem-areas. For FY 2004, systemic reform was "integrated" into Elementary, Secondary, and Informal Education (ESIE) Sub-activity in order to consolidate NSF's pre K–12 activities in one place.[8] At the same time, pipeline programs have shown relative success in increasing the number of PhD degrees awarded to women and minorities over the 1990s.[9] 'Women and minorities in science and engineering' as statistical categories continue to be counted by NSF's technologies of government, such as those used in the creation of the biannual report *Women and Minorities in Science and* Engineering, mandated by the Science and Technology Equal Opportunity Act of 1980. In addition one of the two main evaluation criteria for all research and education proposals submitted to NSF challenge investigators to demonstrate how the proposed activity will "broaden the participation of underrepresented groups (e.g., gender, ethnicity, disability, geographic, etc.)" (National Science Foundation 2003).

## FLEXIBILITY

The first concerted effort to incorporate flexibility in either science or engineering education and to graduate flexible scientists and engineers came from an 1990 NSF/NAE-sponsored workshop entitled "Engineering, Engineers, and Engineering Education in the 21st century." Roland Schmitt—at the time President of Rensselaer Polytechnic Institute, NSB chairman, and the workshop's chairman—questioned the emphasis on engineering sciences in place since the 1960s: "the unanticipated consequences of emphasis on engineering science were to ignore manufacturing, to focus on sophistication of design and features, and less on cost and quality. Some of the engineering education decisions made in the past had detrimental effects on competitiveness. . . . We need to develop a more flexible definition of 'engineers' and 'engineering'"(Schmitt 1990). The NSF responded to this challenge with the funding of the Engineering Education Coalitions (EEC), a program whose main goal was "to produce new structures and fresh approaches affecting all aspects of US undergraduate engineering education, including both curriculum content and significant new instructional delivery systems" (National Science Foundation 1993).[10]

With funding for the Coalitions in place, NSF officials wanted to ensure that this investment resulted in broad systemic changes in engineering education. NSF sponsored two workshops in systemic engineering education reform that were attended by some of the most influential figures in industry, government, and academic with strong interest in reforming engineering education to educate flexible engineers. The first workshop called for "processes for more dramatic change, *enabling curricula to adapt quickly to societal needs, analogous to 'flexible and agile' manufacturing techniques.* Just as we need mechanisms for quickly assembling new programs we need mechanisms for disassembling them when their time is past" (National Science Foundation 1995b, p. 4) (italics mine). Coming from a group committed to systemic reform, this call wanted to ensure that *change* and *adaptation* of curricula and programs became a standard feature of reform. As we will see, by the end of the 1990s the Accreditation Board for Engineering and Technology (ABET) will become the standardizing mechanism to promote change and adaptation in engineering education.

The second workshop justified the need for change in engineering education as follows: "The shift from defense to international competition as a major driver for engineering employment; opportunities offered by intelligent technology to be more creative and 'work smarter'; an expanding social infrastructure that demands a talent for complexity; an eclectic, constantly-changing work environment, calling for astute interpersonal skills; and massively integrated populations placing environment, health, and safety at the front end of design *will require engineers whose intellectual skills include, but extend well beyond, the traditional science-focused preparation that has characterized engineering education since World War II*" (Peden, Ernst et al. 1995) (italics mine). The significance of this call is that for the first time a broad group of educators, corporate officials, NSF administrators, and accreditation evaluators agreed that in order to be flexible for the demands of global competition, engineers' knowledge needed to extend beyond science. As we will see, ABET's accreditation criteria after year 2000 came to reflect this view.

In order to carry out the proposed reforms, workshop attendees were and have been located in positions of great influence at major engineering schools, influential high tech corporations, the NSF, and ABET. When I one of the founding fathers of the Coalitions how the U.S. was going to disseminate this new vision of engineering education based on flexibility, the response was: "That's exactly what the Coalitions program at NSF is aimed to do: cultural change. They want to develop a systemic change of how engineering education is done." What was needed now was a mechanism to encourage or enforce such change.

## ABET 2000

By the mid 1990s, the US government did not have a technology of government that would contribute to solving the needs of a flexible workforce for global competition. The pipeline, as a technology of government, was designed for the needs of the 1980s, i.e., to count and predict the required number of scientists and engineers required to compete with the Japanese. It could not, however, ensure that these scientists and engineers were flexible or that they could operate in an international or multicultural environment. In the 1990s, the problem of governmentality in science and engineering education was framed as a problem of *standards* (systemic reform) and *knowledge and skills* (flexibility) development and dissemination. The pipeline could not address either problem. Educational policymakers understood that the government could not force colleges and universities to adopt standards and to disseminate a particular set of knowledges and skills for global competition. But could they develop non-governmental mechanisms to encourage colleges and universities to adopt standards and disseminate knowledge and skills for flexibility? By the end of the 1990s ABET became the instrument to pressure engineering programs to adopt systemic reform and flexibility.

Given its over-prescription of science content and its "bean-accounting" approach adopted in the 1960s, ABET accreditation process was considered detrimental by champions of educational reform throughout the 1980s and 1990s. According to John Prados, former president of ABET, "In the years following World War II, new engineering programs and the accreditation workload proliferated, engineering education drifted away from its roots in practice, and litigation gained popularity as a way to settle disputes. In response, ABET accreditation became more rigid and rule-bound. . . . The accreditation criteria grew from a few paragraphs . . . to almost thirty pages of fine print containing detailed prescriptions for required courses, credit hour distribution, numbers of faculty, and laboratory improvement plans. The specification-oriented criteria attracted specification-oriented engineers as program evaluators, discouraging those with more flexible views favoring innovation and experimentation. As an active ABET participant in the 1970's and 80's, I fear that I was a contributor the problem!" (Prados 1997).

Prados was hired by NSF to align ABET's accreditation criteria with the new demands for systemic reform and flexibility in engineering education. As NSF's senior advisor involved with the EEC's, Prados wanted to align ABET with the challenges of the new image of the nation under global competition.

In the past, industry and engineering schools criticized ABET for resisting change towards flexibility in the curriculum by trying to protect Cold War criteria which focused on analytic skills obtained through basic and engineering sciences. Now in the 1990s, they wanted ABET to respond to the new needs of the nation. In the words of a senior manager at NSF: "Prados came to NSF to change ABET from an organization that used to be seen as a barrier to the development of engineering education into an organization that becomes the catalyst for new change in engineering education" (interview with author).

Challenged by the new image of the American nation, corporate America demanded a flexible workforce to remain competitive in a tri-polar world. Through ABET companies could ensure the dissemination of these needs into engineering schools. Using Boeing as an example, a NSF high-ranking official explained what corporations tried to accomplish through ABET:

> Boeing is trying to do two things. First of all, to make sure that they have a common understanding within the company of what they're looking for. And then to make sure that the schools understand it. They let this to be known and try to identify some partner schools which are going to be their major target schools for recruitment. . . . And then they will say, "If you turn out graduates who can so this, you can keep being one of our partner schools. If not, we will look else where" . . . Eventually ABET will focus entirely on what you might call "outcome-driven accreditation process" [with] characteristics very similar to the characteristics that Boeing puts out in their list (interview with author).

As we have seen, systemic reform emphasis on outcomes, measurement, and evaluation of student performance, is ensured by a system of policies, governance and resource allocation. Through dissemination of its outcome-driven assessment metrics, ABET could help ensure that companies like Boeing get the kind of student they want from *any* engineering school in the nation. Advocates of flexibility initiatives embraced systemic reform as the mechanism to carry out national dissemination of assessment criteria to ensure proper implementation of new curricula changes such as those developed by the EEC's. The NSF high-ranking official explained how the Coalitions could benefit from systemic reform:

> The Coalitions have the potential, certainly, of making some major contributions. However, they got problems . . . [if] you are trying to generate sustainable systemic reforms, to devise workable strategies, to support diversity and linkage goals, you've got to demonstrate . . . and this is one thing that people are having great difficulty doing: setting up valid metrics to show that the things you are doing are indeed producing the effects that you say you are trying to produce. The sorts of things that you are trying to assess are not simple knowledge kind of things that you can assess on a true false, or multiple-choice test [such as] a standardized test. You

> are trying to assess much higher order skills, and so you tend to move to things like evaluations of portfolios. You are trying to assess whether the students are really developing the skills you want them to. But it's a very complicated process, and all the coalitions are struggling with that (interview with author).

Systemic reform advocates also embraced flexibility in their efforts to educate and train America's workforce for the 21st century. Systemic reform envisioned a fundamental transformation of America's workforce by having an impact on the 95% of students that the pipeline neglected, i.e., those who are left out of science and engineering higher education. When I asked an NSB member who advocated systemic reform, if as part of its grand vision of completely transforming the educational system, systemic reform also aimed at instilling flexibility in the general workforce, the response was:

> Absolutely!! We absolutely have to have the people who are able to respond to the changing requirements of the global marketplace. They have to be able to deal with issues such as customization, working smarter, etc. . . . they are going to have a different job in five years whether they stay on the same job or if they move. . . . The technology basically affects and permeates the structure of all work. . . . Systemic reform holds out a new goal for everyone, a basic level of literacy in science, mathematics, and technology, that gives them a foundation for further life-long learning, and for further training in jobs where that are being affected by science and technology (interview with author).

By the end of the 1990s the push for systemic reform in engineering education to produce the flexible engineers required by global competition transformed ABET into a catalyst for change. Not surprisingly, the new ABET criteria aims at producing both flexible engineers and flexibility in the educational process. Although ABET criteria does not list the word 'flexibility' explicitly in its criteria, the call for engineering programs to graduate flexible engineers should be read from the list of attributes of engineering graduates taken as a whole. After year 2000, ABET requires engineering programs seeking accreditation to show that their graduates posses: "a) an ability to apply knowledge of mathematics, science, and engineering; (b) an ability to design and conduct experiments, as well as to analyze and interpret data; (c) an ability to design a system, component, or process to meet desired needs; (d) an ability to function on multi-disciplinary teams; (e) an ability to identify, formulate, and solve engineering problems; (f) an understanding of professional and ethical responsibility; (g) an ability to communicate effectively; (h) the broad education necessary to understand the impact of engineering solutions in a global and societal context; (i) a recognition of the need for, and an ability to engage in life-long learning; (j) a knowledge of contemporary issues; (k) an ability to use the techniques, skills, and modern engineering tools necessary for engineering practice" (ABET 2002).

### ABET Accreditation: Metaphor for Flexibility

The process of accreditation itself mimics a process of product customization, an inherent part of a flexible system of production and consumption. The requirements of an outcome-based feedback process, as stated in ABET's accreditation criteria, are meant to allow customers (employers of engineering graduates) to provide input on the quality of the product (engineering graduates) to producers (engineering programs) who are supposed to respond to these concerns by incorporating feedback into the curriculum at the appropriate place and time, or 'just-in-time', in order to bring about the desired change in students. According to ABET criteria, each engineering program seeking accreditation must have in place: "(a) detailed published educational objectives that are consistent with the mission of the institution and these criteria; (b) *a process based on the needs of the program's various constituencies* in which the objectives are determined and periodically evaluated; (c) a curriculum and processes that ensure the achievement of these objectives; (d) *a system of ongoing evaluation that demonstrates achievement of these objectives and uses the results to improve the effectiveness of the program*" (ABET 2002) (italics mine). As faculty members from one of the first accredited programs under the new criteria put it, "it is the intent of ABET that engineering education *be shaped by the consumers* of electrical and computer engineering graduates in addition to being *shaped by the producers* of electrical and computer engineering graduates"(Thomas and Alam 2003). So ABET not only encourages the production of flexible engineers but the accreditation process itself represents a flexible production system.[11]

By the end of the 1990s the main elements of systemic reform in engineering education were in place: *engineering programs* committed to change, like those in the Engineering Coalitions, *NSF* as main funding source, *ABET* as enforcement mechanism through accreditation, *ASEE* and other education-oriented societies as disseminators of information, and even prestigious rewards such as the Gordon prize awarded by the National Academy of Engineering as new incentives for educational innovators.

## CONCLUSIONS

In the 1990s an image of the American nation based on global competition between the U.S. and its Asian and European competitors emerged. NSF's mission was redefined around this image of nation that called for internationalization in the creation of knowledge and flexibility in human resources. Challenged by this image, policymakers constructed a discourse where flexibility emerges as the most desirable characteristic for the future techno scientific workforce to have. Flexibility becomes defined as the ability to integrate "just-in-time" skills, including multicultural/national understanding, to create knowledge as well as to identify and solve problems in any global setting.

Under this image and discourse, two new educational initiatives emerged at NSF with the goal of developing the U.S. scientific and technological workforce for the 21st century: systemic reform and flexibility. These initiatives became models that guided education and human resources in science and engineering during the 1990s. Both initiatives tried to redefine the statistical categories that the pipeline created, from gender and race to performance and flexibility. As we have seen, the pipeline studies lost their credibility but the pipeline metaphor lives on. At the same time, ABET accreditation emerges as the mechanism through which systemic reform and flexibility champions want to ensure that flexible engineers graduate from US engineering programs.

As a new image of the American nation threatened by international terrorism emerges in the early 21st century, policymakers have renewed an emphasis on increasing the number of native-born scientists and engineers, particularly by using underrepresented groups like women and minorities, and creating technologies of government to better understand the dynamics of the native S&E workforce abroad and the international S&E in the U.S. This renewed interest in numbers will only reinforce the pipeline metaphor at the same time that the need for flexible skills will make ABET-like accreditation processes more prevalent in science and engineering.

## NOTES

1. Examples of academic representations are Bergsten 1990; Leuenberger and Weisntein 1992; Jackson 1993.

2. Depending on the source, they can be as few as four countries (South Korea, Taiwan, Hong Kong, and Singapore) or as many as seven countries (South Korea, Taiwan, Hong Kong, Singapore, Malaysia, Thailand, and Indonesia) considered part of this emerging economic block. Generally, the first four are granted the label of "little dragons" or "tigers" while Malaysia, Thailand, and Singapore have been assigned the label of "tiger cubs." Japan is still the only country who has earned the full title of "Dragon."

3. The list of witnesses discussing reports and their implications in the future mission of the NSF include: Daniel Nathans, from the President Council of Advisors on Science and Technology (PCAST) discussing Renewing the Promise: Research-Intensive Universities and the Nation; Guyford Stever, Chairman of the Carnegie Commission Task Force discussing its report on Enabling the Future: Linking Science and Technology to Society Goals; John Wiley, from the Government-University-Industry

Research Roundtable's (GUIRR) Working Group, discussing the report Fateful Choices: The Future of the U.S. Academic Research Enterprise.

4. The most influential work of these comparisons was conducted by MIT's Commission on Industrial Productivity about developing the human resources for technological advance and competitiveness. This work was eventually incorporated into Dertouzos, Lester et al. 1989. During my tenure at NSF from 1989 to 1991, this book was held up as the bible of competitiveness through flexibility and was very influential in the conceptualization of the Engineering Education Coalitions.

5. For further details of these interconnected aspects of systemic reform, see Smith and O'Day 1991. For a comprehensive account of how systemic reform developed out of the Goals 2000 summit, see Riley 1995.

6. Actually, NSF's education budget went from 16% of NSF's total in 1993 to 19% in 1994 and has remained at a steady 18% of NSF's total to the present.

7. Evidence of this was the emergence of the new Indicators of Science and Mathematics Education at NSF under a separate cover from the traditional Science and Engineering Indicators. These new indicators "address, in more detail than any other report, the progress made towards the U.S. national goal of ranking first in the world in mathematics and science by the year 2000" (National Science Foundation 1993b). These indicators were short lived, disappearing one year after their first publication.

8. From FY 1999 to 2003, the budget for NSF's Educational Systemic Reform Subactivity decreased from $116 million (or 17% of NSF's total education budget) to $40 million (or 4.4% of NSF's total education budget).

9. Between 1990 and 1999, the percentage of PhDs awarded to women in science and engineering increased from 28% to 35% and to minorities from 14% to 22%. (NSF 2002b).

10. By the late 1990s NSF had spent more than $200 million in funding the following Coalitions: Engineering Coalition of Schools for Excellence in Education and Leadership (ECSEL) (1990–94), Synthesis (1990–94), Gateway (1992–96), Southeastern University and College Coalition for Engineering Education (SUCCEED) (1992–96), Foundation (1993–97), and Greenfield (1993–97). Not surprisingly the main three foci of the ECCs were manufacturing, design, and curricular integration.

11. Some engineering programs have reported that the most difficult part of the new accreditation process is closing the feedback loop or materializing outcome assessment data from customers into curricular improvements. In short, product customization has proven to be difficult to accomplish. Apparently, educational institutions are too rigid and slow to respond. Realizing this, some schools have created full-time assessment centers "to provide sustainability to the assessment process" (Burke and Rainey 2003). If the pipeline created an industry in the late 1980s, outcome assessment is on its way to create an assessment industry in the near future as evidence by the inclusion of outcome assessment as requirements in NSF's criteria to fund educational projects, the establishment of the Council for Higher Education Accreditation (CHEA) established in 1996, and the growing number of outcome-based assessment conferences and workshops being held by most major organizations in higher education.

## *Chapter Six*

# Conclusion

Policymaking in science and engineering (S&E) education at the NSF is a history of national culture. Throughout this account, we have seen how cultural images of nation are constructed and invoked by social actors in their effort to make policies and programs to educate and train scientists and engineers for specific national purposes. Hence, what we understand as "scientist" and "engineer" at any particular time is inseparable from these images of nation. Nowhere is this cultural relationship between nation, policymakers, and S&E education more evident than around the NSF.

In the last five decades, we have seen how the dominant image of the American nation shifted from being under threat by Soviet communism in the 1960s, domestic social and environmental problems in the 1970s, Japanese technological competitiveness in 1980s, global competition in the 1990s, and more recently by terrorism. Challenged by these images, specific actors and groups set the limits of discourse and called for specific kinds of science and technology to save the nation: scientific academists called for *elite science* in the 1960s; science critics and reformers advocated *appropriate science and technology* in the 1970s; technophiles endorsed *technology for economic competitiveness* in the 1980s; technophiles and educational reformers are still calling for *flexible technoscience for global competition* since the 1990s. Nowadays, we are witnessing the emergence of a parallel discourse of *patriotic science and technology* to combat terrorism.

Throughout its 55-year history, the NSF's S&E education programs have increasingly become a site where these actors and groups struggle to address the challenges posed by images of nation in terms of specific kinds of scientists and engineers: *high quality scientists* for a nation under threat by Soviet science, *mobile and interdisciplinary scientists and engineers* for a nation under threat by its own domestic problems, *lots of engineers* for a nation under threat by Japan's technological success, *flexible scientists and engineers* for a nation engaged in global competition, and *patriotic scientists and engineers* for a nation under the threat of terrorism. Throughout the process of developing educational programs and projection models for scientists and engineers, underrepresented groups, such as women and minorities, have found and created space for participation in science and engineering. This is what happened in the 1970s when programs for *mobility and interdisciplinarity* provided the opportunity for women and minorities to claim their participation in the solution of national problems. With the help of powerful actors like Senator Edward Kennedy, they pushed NSF to recognize them as categories worthy of budget allocations in S&E education. In the 1980s, when technophiles at NSF defined the problem of human resources for competitiveness as a matter of *numbers* and recognized the demographic potential of underrepresented groups, women and minorities successfully argued that they could contribute numerically to the solution of this problem.

Groups involved in S&E education policymaking encountered different political challenges in the last five decades. In the 1960s scientific academists received no resistance to their efforts in setting the limits of discourse and subsequent policies because most groups involved in S&E policymaking were aligned behind the image of the American nation under the threat of Soviet communism. However, in the 1970s scientific academists endured resistance and criticism from groups and actors seeking a different kind of science and technology. In the 1980s, technophiles did not experience major resistance as all groups in S&E policymaking jumped on the bandwagon of economic competitiveness. In the 1990s, two groups struggled to shape S&E education: supporters of the pipeline and those of systemic reform.

We have also seen how in order to project the production of scientists and engineers for specific national needs, in both appropriate numbers and characteristics, NSF has redefined its projection models according to the challenges posed by images of nation. In the 1960s "manpower" models to predict the required number of scientists and engineers changed from *free-market* models in the early 1950s to those based on *fixed-coefficients* after Sputnik. In the 1970s, after long resistance from its administrators, NSF incorporated short-lived *flexible-coefficient* model to address the needs of a nation threatened by domestic social and environmental problems. In the 1980s, NSF created and disseminated a *demographics-based* model (pipeline) in response to the threat by Japanese competitiveness.

## WHERE ARE WE NOW?

In 2005, two images of nation challenge policymakers and educators in S&E education. First, an image of a nation under the threat of global competition continues to challenge them to attract and retain foreign scientists and engineers, as nations around the world retain more of their own scientists and engineers instead of losing them to the U.S. The European Union has set the goal of becoming the most competitive knowledge-based economy by 2010, and countries like China and India, who traditionally supply the U.S. with the largest number of foreign scientists and engineers, are creating more welcoming domestic environments to prevent brain drain. At the same time, new immigration rules and hurdles for foreign students and scholars have been set in place since 9/11, making the U.S. not as welcoming as other countries. By fall 2004, headlines in science- and higher-education media read "Foreign Enrollments at American Universities Drop for the First Time in 32 Years" (Bollag 2004a) and "Wanted: Foreign Students" (Bollag 2004b). Challenged by this image of nation, U.S. policymakers and educators try to address the needs of S&E workforce by recruiting from underrepresented groups in science and engineering. Recognizing this challenge, the NSB in its latest report on S&E workforce concludes:

> Global competition for S&E talent is intensifying, such that the US may not be able to rely on the international S&E labor market to fill unmet skill needs. . . . The number of native-born S&E graduates entering the workforce is likely to decline unless the Nation intervenes to improve success in educating S&E students from all demographic groups, especially those that have been underrepresented in S&E careers (National Science Board 2003, p. 1).

At the same time, an image of the American nation under the threat of terrorism also challenges policymakers and educators to tighten U.S. borders to foreign students and scholars and rely more on domestic S&E workforce. Aware of the negative consequences of the war on terrorism on immigration and the increasing competition from other countries for brainpower, U.S. policymakers now understand that the U.S. cannot depend on foreign scientists as it used to. Their response to this challenge is to continue feeding the U.S. S&E pipeline, particularly from underrepresented groups in science and engineering. Recognizing the challenges posed by this second image, the NRC's Committee on Science and Technology for Countering Terrorism claims that:

> Indeed, America's historical strength in science and engineering is perhaps its most critical asset in countering terrorism without degrading our quality of life.

. . . The nation's ability to perform the needed short- and long-term research and development rests fundamentally on a strong scientific and engineering workforce. Here there is cause for concern, as the number of American students interested in science and engineering is declining, as is support for physical science and engineering research. . . . If the number of qualified foreign students decline, the need to reverse this trend will become more urgent (National Research Council 2002, p. 23).

## The Pipeline Lives

In spite of the unfulfilled crisis predicted by the NSF pipeline models in the late 1980s, the pipeline metaphor stuck because both co-existing images challenge policymakers and educators to become less dependent on foreign scientists and engineers and to recruit from U.S. underrepresented groups. Articles continue to appear in U.S. main science magazines and journals with titles such as "The Pipeline: Still Leaking" (Goodchild 2004) and "Patching a leaky pipeline" (Smaglik 2004). Without using the word "pipeline", NSF's biannual publication, *Women, Minorities, and Persons with Disabilities in Science and Engineering*, continues to present its data according to the pipeline metaphor. Beginning with U.S. population, the data is displayed in a sequence, from potential S&E students, to enrollments, attrition rates, undergraduate degrees, graduate enrollments, graduate degrees, post-docs, and, finally, S&E employment (National Science Foundation 2004). And NSB's most recent comprehensive report on S&E workforce, *The Science and Engineering Workforce: Realizing America's Talent*, reaffirms a commitment to the pipeline when it states that "lacking reliable tools for policy and strategies affecting the future S&E workforce, the consensus strategy is to attract more talented undergraduates to science and engineering majors in areas of need and encourage them to continue on to graduate school—particularly undergraduates from groups who have been underrepresented in natural science and engineering"(National Science Board 2003, p. 29).[1] To support this argument, NSB report cites a number of high-profile reports which rely heavily on the pipeline metaphor (National Science and Technology Council 2000; Jackson 2002). One of these reports states, "The U.S. lengthy economic boom has hidden the fact that there is not enough technical talent in the pipeline to replace enough of the skilled labor responsible for our country's prosperity" (Jackson 2002, p. 2). I predict that in the near-term future the pipeline will continue to be the metaphor that inspires educators and policymakers to visualize the movement of S&E students, propose solutions to recruitment and retention problems, and project shortfalls in the S&E workforce.

## Need for New Models

This over-reliance on the pipeline metaphor, particularly after the unfulfilled crises predicted by pipeline-inspired models in the early 1990s, has led some to call for new models but to no avail. The NSF sponsored a workshop on *Improving Models of Forecasting Demand and Supply for Doctoral Scientists and Engineers* (March, 1998) which concluded that:

> Interest in predicting demand and supply for doctoral scientists and engineers began in the 1950s, and since that time there have been repeated efforts to forecast impending shortages or surpluses. As the importance of science and engineering has increased in relation to the American economy, so has the need for indicators of the adequacy of future demand and supply for scientific and engineering personnel. This need, however, has not been met by databased forecasting models, and accurate forecasts have not been produced (National Research Council 2000, p. 1).

Five years after the NSF workshop, the need for forecasting models remained unfulfilled as acknowledged in NSB's latest S&E workforce report:

> Over the long term, there is a need to develop a quantitative, dynamic model of the global S&E workforce with respect to skills, mobility, occupational and geographic migration and demographic characteristics, and to understand the impacts of the global workforce on US science and engineering, especially the impacts of temporary workers and international students in S&E fields(National Science Board 2003, p. 29).

There is a need to move beyond the pipeline towards a model of the global S&E workforce now that there is a clear realization that the migration and movement of scientists and engineers are ongoing and dynamic transnational phenomena. Yet we seem to be stuck with the pipeline, as metaphor and inspiration for models, in spite of its embarrassing predictions in the early 1980s and limitations as a national model. Why are shortage claims, based on pipeline-inspired models, so persistent despite much evidence to the contrary? Expanding on economist Eric Weinstein's argument "that the real intent of some of those involved in the 1980s 'shortfall' alarms from NSF may have been to limit wage increase for Ph.D. scientists," Michael Titelbaum argued that shortage crises serve group and institutional interests of all involved in higher education:

> *Universities* want to fill their classrooms with undergraduates who pay their fees and finance their research with external funding, and to do so recruit graduate students and postdoctoral fellows to teach undergraduates and to staff their research laboratories. *Government science-funding agencies* may find rising

> wages problematic insofar as they result in increased costs for research. Meanwhile, *companies* want to hire employees with appropriate skills and backgrounds at remuneration rates that allow them to compete with other firms that recruit lower-wage employees from less affluent countries. If company recruiters find large numbers of foreign students in U.S. graduate science and engineering programs, they feel they should be able to hire such non-citizens without large costs or lengthy delays. Finally, *immigration lawyers* want to increase demand for their billable services, and especially demand from the more lucrative clients such as would-be employers of skilled foreign workers. . . . None of these groups is seeking to do harm to anyone. Each finds itself operating in response to incentives that are not entirely of its own making. But a broad commonality of interests exists among these disparate groups in propagating the idea of a "shortage" of native-born scientists and engineers (Teitelbaum 2003, p. 53) (italics mine).

Group and institutional interests might explain somewhat the endurance of the pipeline from the mid-1980s to the present. However, group interests by themselves cannot explain changes in S&E policymaking after historical events like Sputnik, or the participation and exclusion of certain kinds of groups and emergence of projection models at particular times in U.S. history. The emergence, existence, persistence, and demise of policies and models are better explained by the connection between images of nation, policymaking institutions, groups, and models. As we have seen, models have been developed by actors challenged by a specific image of nation at a particular time. As images of nation change, models change. So the pipeline stuck because, in addition to serving institutional and group interests for the time being, it addresses the challenges posed by the two co-existing images of America under the threat of global competition and terrorism. We should not expect the emergence of a new metaphor to inspire new models or the wide acceptance of new models until the existing images of nation change. Since models are reflections of how policymakers and modelers see the nation, a dynamic model of the global S&E workforce, like the one desired in the NSB report, is unlikely to emerge as a credible source of knowledge until we are challenged by an image of nation integrated into a global community.

## ANALYTIC IMPLICATIONS

### S&E Education

As scholars in fields like science and technology studies (STS), political science, public administration, policy studies, sociology of education, and educational research read this cultural and historical account of policymaking, they might encounter a number of conceptual and methodological challenges that hopefully will motivate future research projects, particularly on the relationship between S&E education and workforce and the nation-state. Throughout this book, the focus has been on the responses of policymakers and educational reformers in the last five decades to the challenges posed by images of the nation. However, many important questions remain unexplored. For example, how have different educational institutions (e.g., private vs. public, or research- vs. teaching-oriented), S&E educators (e.g., college-level vs. elementary) and students (e.g., life sciences vs. engineering vs. liberal arts) reacted to the challenges posed by different images of nation? How has this relationship between nation-state and S&E education and workforce been articulated in other countries, for example, where central governments control higher education vs. those where control is decentralized and federalized?

Here I would like to point to the benefits that other traditions of research can derive from taking on questions of S&E education. First of all, S&E education has large constituencies, as shown by membership in associations like the UK's Association for Science Education and the American Society for Engineering Education (ASEE). In 1999, there were almost 400,000 science and engineering graduates in the U.S., and between 3 to 10 million people working as scientists or engineers, depending on the definition being used (National Science Board 2004). These are significant populations that deserve analytical attention. Second, S&E education and workforce enjoy privileged positions in strategies and policies in most countries wanting to enhance their competitiveness. In 1998, NSF sponsored a workshop aimed at finding out what nations in Europe, Asia, and the Americas were doing in S&E education and workforce policy. According to the workshop report, "as the world's countries recast themselves as 'knowledge-base' economies and build up 'national innovation systems,' interest in S&E education is increasing around the globe, occasioning a reexamination of its aim and structure" (National Science Foundation 2000, p. 1). Since the late 1990s, ministers of science and technology from European and Asian countries have taken the time to inform the S&E community at large of the importance of S&E education for national development and competitiveness (Allègre 1998; Merkel 1998; 1998).

Important research questions about these large and increasingly influential groups remain unexplored. For example, beyond pedagogical and organizational factors involved in students decisions to enter or leave science and engineering (see Seymour and Hewitt 2000), do students make decisions to enroll or leave in response to challenges posed by images of nation found in the media, official discourse, or in the language used by their academic counselors and relatives? Do students' responses to the images change as they move through their career paths? Do members or groups of the S&E workforce migrate to other research areas or countries as a response to challenges from images of nation? If so, what happens to these S&E migrants when they enter foreign national spaces and are challenged by other images of nation different than those found in their home country?

### Models and Projections

Foucauldian analyses of educational, economic, and clinical practices and institutions have shown us how the modern nation-state is constituted by technologies of government (Ball 1990; Burchell, Gordon et al. 1991). However, these analyses have overlooked practices and institutions aimed at counting, predicting, and educating the scientists and engineers that a modern nation-state needs to address its problems. At the same time, population forecasters have been preoccupied with the use and accuracy of forecasting models (Ahlburg and Lutz 1998) but have overlooked the influence of images of nation on these models. The relationship between nation-state and forecasting models remain understudied, even though, as seen through the history of the S&E pipeline, it has significant implications for society at large.

Based on current pipeline-inspired and future models and forecasts, upcoming generations of high school graduates will make career decisions, parents will make significant financial commitments, colleges and universities will make long-term strategic decisions, high-tech corporations dependent on scientific and engineering personnel will make investment and relocation decisions, and governments will make immigration and R&D allocation decisions. So as old forecasting models are used or new ones are proposed, we need to explore the assumptions behind these models and how much their development is influenced by past, existing, or emerging images of nation. Would the new models continue to serve exclusively the interests of groups and institutions as Titelbaum argues? Or would models, due to their national origins, privilege those groups and institutions rooted in the nation-state (e.g., governments, universities, ethnic groups recognized by census) over those that transcend the nation-state (e.g., international organizations, multinational corporations, migrant groups)? Do other cultural assumptions, beyond images of nation, influence the construction of projection models? If so, what are they?

## POLICY RECOMMENDATIONS

With each successive set of policies, programs, and models for S&E education and human resources, was there anything learned from the previous ones? In spite of the changes in images of nation, is there any continuity in what is learned from one generation of policymakers to the next? Here are some of the main lessons learned from this historical account.

### Images Count

First, dominant images of the nation constrain to a significant extent the limits of discourse (i.e., what can and cannot be said in policy) and the limits of appropriation (i.e., who can and cannot participate in policymaking). Second, alignment with a dominant image of nation makes policy proposals more likely to succeed. The clear alignment of scientific academists in the 1960s and technophiles in the 1980s with the dominant images of nation ensured to a large extent the success of their policies and programs. Third, resistance to a dominant image of nation makes policy proposals likely to fail. For example, we saw how policy proposals by advocates for women and minorities in science and engineering failed during the early 1980s until they were aligned with the image of America under the threat of Japanese competitiveness. Fourth, images do not exist by themselves. An interplay between historical circumstances and social actors create the meaning of these images and ensure their continuity. We have seen how under different historic circumstances, the media and powerful groups interacted closely, gave meaning to, and disseminated images of nation. For example, scientific academism in the 1960s and the technophiles in the 1980s played significant roles in shaping images of American nation under different threats by effectively using popular and scientific media.

*Recommendations.* Contemporary educational reformers in S&E education might consider alignment with current and future images of nation in order to promote their policy proposals and make them more likely to succeed. However, in order to avoid complete co-optation by images with which they do not fully agree, reformers need to get involved in the re-articulation and dissemination of meaning of these images. For example, some might find it preferable to align their proposals with the image of nation under the threat of global competition, rather than with the one under the threat of terrorism, and shape the responses to this image in terms of *high quality* rather than *low cost*. As an answer to global competition, reformers could propose to make the U.S. more competitive by attracting foreign and domestic students into high-quality S&E programs in spite of higher costs. Unlike most countries, U.S. higher education enjoys the benefits of having S&E programs in close proximity to the humanities and social sciences (HU/SS), an evolving cadre of educational researchers and institutions, and few but clear examples of S&E programs relevant to social and environmental problems (see Lucena 2003 for examples). Hence, reformers have a unique opportunity to develop high-quality S&E programs in terms of critical thinking, learning assessment, and relevance to social and environmental problems by properly integrating S&E education with the HU/SS, widely implementing effective pedagogies based on sound educational research, and developing and adopting social- and environmental-relevance criteria in the selection of research and educational topics. These programs will be more attractive and competitive than those found anywhere else.

## Access to Power

In setting the limits of appropriation—i.e., what individual, groups, classes, or organizations have access to a particular kind of discourse—we have learned a number of important lessons. First, limits are largely defined by those with access to political power. For example, scientific academists in the 1960s and technophiles in the 1970s were able to set the limits of discourse and direct policy because of their access to the White House, corporate America, and America's most powerful higher education institutions. Second, groups with less power rely on alliances with those in positions of power in order to enter and influence the limits of discourse. For example, the alliances created between the groups representing the interests of Black Americans in science and powerful congressmen like Kennedy in the 1970s, and between groups representing women and minorities and powerful federal officials like Erich Bloch in the 1980s were successful in bringing about policies and legislation to open access in science and engineering to women and minorities.

*Recommendations.* The success of current and future educational reforms will depend, to a large extent, on the access that reformers get to the Executive branch of government, their presence in top-administrative positions in elite institutions of higher education, and the support that they obtain from corporate America. For example, educational reformers who are committed to improving quality of science and engineering education, but resist the image of the US under the threat of terrorism, might have to align themselves with muti-national corporations committed to excellence and quality in their S&E workforce. Companies like Boeing, Hewlett Packard, and Microsoft and philanthropic organizations like the Carnegie Foundation have shown commitment to education reform. The Federal Government presents opportunities for reformers to intervene in S&E education policy through programs like the Congressional Fellows. Educational reformers could further their goals by intervening, for example, in advisory committees and review panels of corporate and governmental educational programs, and help shape their agendas and budgets.

## Location Matters

Location inside the bureaucracy, at all levels, matters. With support from NSF directors and NSB members, actors inside NSF divisions had significant success on how their technologies of government influenced policy decisions and budget allocations for many years. For example, in the 1960s, scientific academism had strong advocates at different levels of NSF's bureaucracy who ensured that discourse would actually materialize in programs and budgets, like Harry Kelly those who implemented fixed-coefficient models in the division of scientific personnel and education (SPE) (see footnote 17 in chapter 2). Reformers located in the NSB influenced the development of short-lived 'manpower' model of the 1970s and systemic reform initiatives in the 1990s. Technophiles inside NSF's PRA division had great deal of influence with the development of the pipeline in the 1980s.

*Recommendations.* Educational reformers have made some progress by locating themselves within S&E bureaucracies, as shown by those working at the Center for the Advancement for Science and Engineering Education (CASEE) at the National Academy of Engineering and at NSF's Directorate for Education and Human Resources. But key locations remain to be influenced throughout NSF directorates and elsewhere in the federal, state, and local governments. Upcoming generations of reformers might want to consider job opportunities in research, analysis, and policy within the NSF or in academic administration of universities with significant influence in S&E education. [2]

## Are S&E Workforce Crises Inevitable?

As responses to the challenges posed by images of nation, models changed when images changed. This might help explain why new models always predict a "shortage" or "shortfall" from the previous one. As assumptions and scenarios about new needs of the nation are built in new models, new and unfulfilled workforce needs are likely to emerge. Hence, as long as models are constructed to address the challenges of the times, they will continue to predict crises.

*Recommendations*. All of us who are involved in helping governments, institutions, and individuals make decisions based on S&E workforce models and projections need to educate them about the strengths and limitations, including national- and time-specific biases in the assumptions built into these models and projections. For example, my S&E students struggle interpreting the meaning of recent headlines (e.g., declaring "shortages" or calling for foreign students) for the future careers. It is my responsibility to show them the assumptions behind these claims to help them make educated choices.

## Under-representation in S&E and the Need for Reform

As we saw in chapter 4, most of the existing and near-future recruitment and retention initiatives continue to be inspired by the pipeline. Yet the pipeline seems to have outlived its effectiveness in increasing representation. After millions of dollars spent in recruitment and retention initiatives in the last 20 years, we seem to have reached plateaus in the representation of women and minorities in science and engineering. For example, the representation of women in engineering programs has not surpassed 20%, and of Blacks and Hispanics in engineering has not been higher than 6% after two decades of pipeline-inspired initiatives.

More recently, educational reformers have developed a number of pedagogical reform initiatives focused on teaching, learning, and assessment. This reform movement has been labeled "scientific teaching" (Handelsman, Ebert-May et al. 2004). It is too early to tell the wide impact of these initiatives on increasing representation. However, one could predict a slight positive impact because most initiatives are based on the assumption that different people learn differently (see, for example, Springer, Stanne et al. 1999). Accordingly, most educational reformers have adjusted their pedagogies to different learning styles. One could predict that these changes will improve retention at the institutional level and eventually at the national level (Felder, Woods et al. 2000).

*Recommendations.* Beyond pedagogical reforms, S&E educators need to focus on curricular and programmatic reform and development. After an exhaustive review of recent engineering education reform initiatives, Ilene Busch-Vishniac and Jeffrey Jaroz concluded that

> There is certainly ample evidence that the undergraduate student population in engineering is not terribly diverse, with the situation worsening in spite of ongoing attempts at improvement . . . we believe that a revolutionary change is needed—one that addresses the root cause: an unattractive, unresponsive, and culturally biased curriculum—rather than an easing of symptoms through numerical targets achieved by any means possible (Busch-Vishniac and Jarosz 2004, p. 276).

There is great need for meaningful revolutionary educational reform, one that aligns science and engineering education with the needs of society, the environment, and the desires of groups that have remained excluded from science and engineering. However, reform advocates need to increase their effectiveness by following some of the lessons described above. The key challenge for the next generation of S&E educational reformers is how to respond to existing and new images of the nation, shape the meaning of these images, secure access to power, form alliances, and strategically locate themselves inside bureaucracies while helping S&E students learn better and focus their knowledge and skills in addressing meaningful social and environmental problems.

## NOTES

1. Present areas of need in science and engineering are determined by the difference between the data collected by the Bureau of Labor Statistics (BLS) National Industry Occupation Employment Matrix (NIOEM) and NSF's Scientists and Engineers Statistical Data Systems (SESTAT). "The NIOEM data give a broad view of the *demand* reported by establishments in the U.S. (What are the jobs that are available?); the SESTAT data give a more detailed view of the *supply* side reported by bachelor's and above scientists and engineers employed in the labor force (Who are the persons available to fill those jobs?)" (National Science Foundation 1999) (italics mine).

2. At the time of this writing, there were two Senior Science Resource Analysts positions available at NSF's SRS division to develop "quantitatively based analyses on Asian S&E policies, patterns and trends, often in comparison with the United States and the European Union."

# Bibliography

"America vs. the World." *Fortune*, 9 March 1992.

"Apollo's moon mission: Here are the results." *U.S News & World Report,* 4 August 1969b.

"Can U.S. Hold Its Lead Over Soviets In Science Race?" *U.S. News and World Report*, 1 October 1984, 51–54.

"China Plans Major Shake-up of Academy." *Nature*, July 2 1998, 7.

"Comparison of All Russian Sciences." *Newsweek*, 11 November 1957, 108–22.

"Crisis in Education." *Life*, March 24 1958, 26–33.

"Defense Priority: High-Technology Weapons." 1981.

"Education for All American Youth." Washington, D.C.: National Education Association, 1944.

"Educators Upset by Soviet Stroke." *New York Times*, 11 October 1957c.

"Educators Urge High School Shift." *New York Times*, 31 October 1957d.

"Educators Urge Wakened Nation." *New York Times*, 9 November 1957e.

"The feat that shook the earth." *Life* 1957b.

"The Future: Made in Japan." *Science Digest* 1981, 65–77.

"How U.S. Beat Russia to the Moon." *U.S. News and World Report* 1969a.

"Innovation: Japan Races Ahead As U.S. Falters." *Science*, 14 November 1980, 751–54.

"It's Austerity Time for Basic Science." *Business Week*, 3 February 1973, 47–49.

"Japan's Strategy for the '80s." *Business Week*, 14 December 1981, 39–119.

"Man on the Moon: Mixed Emotions." *Science News*, 26 July 1969, 71.

"Man's Awesome Adventure." *Newsweek*, 14 October 1957, 37–41.

"The Painful Struggle for Relevancy." *Business Week*, 2 January 1971, 38–9.

"The Productivity Crisis—Can America Renew Its Economic Promise?" *Newsweek*, 8 September 1980, 50–9.

"Reaching Beyond the Rational." *Time*, 23 April 1973, 83–86.

"Satellite Called Spur to Education." *New York Times*, 12 October 1957f.

"Science: America's Struggle to Stay Ahead." *U.S. News and World Report*, 15 September 1980, 52–55.

"Soviet satellite sends U.S. into a tizzy." *Life* 1957a.

"Why Did U.S. Lose the Race? Critics Speak Up." *Life*, Nov 1957, 22–3.

"World War of Science. . . . How we are mobilizing to win it." *Newsweek*, 18 November 1957, 37–8.

Abelson, Philip. "Science and Immediate Social Goals." *Science* 1970, 1.

ABET. *Criteria for Accrediting Engineering Programs—Effective for Evaluations During the2003–2004 Accreditation Cycle*. 3rd ed. Baltimore: ABET, 2002.

Aerospace Education Foundation. "America's Next Crisis: The Shortfall in Technical Manpower." Arlington, Virginia, 1989.

Ahlburg, Dennis A., and Wolfgang Lutz. "Introduction: The Need to Rethink Approaches to Population Forecasts." *Population and Development Review* 24 (1998): 1–14.

Ailes, Catherine , and Francis Rushing. "The Science Race: Training and Utilization of Scientists and Engineers, US and USSR." Menlo Park: Stanford Research Institute, 1982.

Allègre, Claude. "French Strategy for Science Education. Editorial by the Minister of Education, Research and Technology." *Science*, July 24 1998, 515.

Alzinga, Aant and Andrew Jamison. "Changing Policy Agendas in Science and Technology. In Handbook of Science and Technology Studies." edited by Sheila Jasanoff et al., 572–97. London: Sage, 1995.

Anderson, Benice. "How can middle schools get minority females in the math/science pipeline?" *The Education Digest* 59 (1993): 39–42.

Astin, Alexander W et al. *The American Freshman: Twenty Year Trends*. Los Angeles, CA: UCLA, 1987.

Atkinson, Richard. "Supply and Demand For Scientists and Engineers: A National Crisis in the Making." *Science*, 27 April 1990, 425–32.

Averch, Harvey. *A Strategic Analysis of Science and Technology Policy*. Baltimore: John Hopkins, 1985.

Bailey, R. "Black students and the mathematics, science, and technology pipeline: turning the trickle into a flood." *The Journal of Negro Education* 59 (1990): 239–516.

Ball, Stephen J., ed. *Foucault and Education*. London: Routledge, 1990.

Barlett, Thomas. "Degrees of security: Colleges start programs to train students for jobs in homeland defense." *The Chronicle of Higher Education* 2003, A24.

Benjamin, Milton. "Sputnik Plus 20: The U.S. On Top." *Newsweek*, 10 October 1977, 52–3.

Bergsten, C. Fred. "Japan and the United States in the New World Economy: Tripolarity." *Vital Speeches of the Day* (1990): 653–57.

Bernes, Esmeralda. "Getting minorities into Ph.D. pipeline requires active and early intervention, say experts." *Black Issues in Higher Education* 9, no. July 16 (1992): 8–9.

Bestor, Arthur E. "Educational Wastelands: The Retreat from Learning in Our Public Schools." Urbana, Ill., 1953.

Bimber, Bruce, and David Guston. "Politics by the Same Means: Government and Science in the United States." In *Handbook of Science and Technology Studies*, edited by Sheila Jasanoff et al., 554–71. London: Sage, 1995.

Blank, David M. , and George J. Stingler. *The Demand and Supply of Scientific Personnel*. New York: National Bureau of Economic Research, 1957.

Bloch, Eric. "Engineering Education in the 1990s." Paper presented at Purdue University EE Distinguished Lecture, April 6, 1989.

Bollag, Burton. "Foreign Enrollments at American Universities Drop for the First Time in 32 Years." *The Chronicle of Higher Education*, November 10 2004, A1.

———. "Wanted: Foreign Students." *The Chronicle of Higher Education*, October 8 2004, A37.

Branscomb, Lewis M. *Confessions of a Technophile*. Woodbury, NY: American Institute of Physics, 1995.

———, ed. *Empowering Technology: Implementing a U.S. Strategy*. Cambridge: MIT Press, 1993.

Burchell, Graham. "Liberal government and techniques of the self." *Economy and Society* 22, no. 3 (1993): 267–82.

Burchell, Graham, Colin Gordon, and Peter Miller, eds. *The Foucault Effect: Studies in Governmentality*. Chicago: University of Chicago Press, 1991.

Burke, Kena, and Paul E. Rainey. "Cal Poly Engineering Assessment Center—How It Works." Paper presented at the 2003 ASEE Annual Conference, Nashville, TN 2003.

Busch-Vishniac, Ilene J., and Jeffrey P. Jarosz. "Can diversity in the undergraduate engineering population be enhanced through curricular change?" *Journal of Women and Minorities in Science and Engineering* 10 (2004): 255–81.

Business-Higher Education Forum. *America's Competitive Challenge: The Need for a National Response*: Business-Higher Education Forum, 1983.

———. *Engineering Manpower and education: Foundation for Future Competitiveness*: Business-Higher Education Forum, 1982.

Cass, James. "In the Service of Man." *Saturday Review*, 21 March 1970.

Clowse, Barbara. *Brainpower for the Cold War: The Sputnik Crisis and National Defense Education Act of 1958*. Westport: Greenwood Press, 1981.

Colwell, Rita R. "Science as Patriotism." Paper presented at the Annual Meeting of the Universities Research Association, Washington, D.C. 2002.

Conant, James B. *Modern Science and Modern Man*. New York: McGraw, 1952.

———, ed. *General Education in a Free Society*. Cambridge: Harvard University Press, 1945.

Cooper, Edith Fairman. "U.S. Science and Engineering Education and Manpower." Washington, D.C.: Library of Congress, 1983.

Cooper, Gail. "Technological Leadership and International Competitiveness: A Comparative Approach." In *Competitiveness and American Society*, edited by Steven L. Goldman, 206–28. Bethlehem: Lehigh University Press, 1993.

Cooper, Kenneth J. "Study Sees Shortfall of Graduates in Technical Fields." *Washington Post*, 19 December 1989, A-6.

Cruikshank, Barbara. "Revolutions within: self-government and self-stem." *Economy and Society* 22, no. 3 (1993): 327–44.

Dauffenbach, R.C. and Fiorito. "Projections of Supply of Scientists and Engineers to Meet Defense and Nondefense Requirements 1981–87." Oklahoma State University: Contractor report to the NSF, 1983.

De Witt, Nicholas. *Education and Professional Employment in the USSR*. Washington, D.C.: National Science Foundation, 1961.

———. *Soviet Professional Manpower: Its Education Training and Supply*. Washington, D.C: National Science Foundation, 1955.

Dertouzos, Michael L., Richard K. Lester, and Robert M. Solow. *Made In America: Regaining the Productive Edge*. Cambridge: MIT Press, 1989.

Dow, Peter B. *Schoolhouse Politics: Lessons from the Sputnik Era*. Cambridge: Harvard University Press, 1991.

Downey, Gary. *The Machine in Me: An Anthropologists Sits Among Computer Engineers*. New York: Routledge, 1998.

———. "Steering Technology Development Through Computer-Aided Design." In *Managing Technology in Society*, edited by Johan Schot. London: Pinter Publishers, 1995a.

———. "The World of Industry-University-Government: Reimagining R&D as America." In *Technoscientific Imaginaries*, edited by George E. Marcus. Chicago: University of Chicago Press, 1995b.

Downey, Gary , and Juan C. Lucena. "Engineering Selves." In *Cyborgs and Citadels: Anthropological Interventions in Emerging Sciences and Technologies*, edited by Gary L. Downey et al., 117–42. Seattle: University of Washington Press, 1996.

Downey, Gary L., and Juan C. Lucena. "Knowledge and Professional Identity in Engineering: Code-Switching and the Metrics of Progress." *History and Technology* 20, no. 4 (2004): 393–420.

Dupree, A. H. "A New Rationale for Science." *Saturday Review*, February 7 1970, 55–57.

Eisenhower, Dwight. "Presidential address on "Our Future Security"." Oklahoma City, 1957.

Emens, John R. "Education Begets Education: The G.I. Bill Twenty Years Later." *American Education* 1 (1965): 11–13.

Engardio, Pete. "21st Century Capitalism: The New Global Workforce." *Business Week* 1994, 110–34.

England, J. Merton. *A Patron for Pure Science: The National Science Foundation's Formative Years, 1945–57*. Washington, D.C.: NSF, 1982.

Feinberg, Walter. *Japan and the Pursuit of a New American Identity: Work and Education in a Multicultural Age*. New York: Routledge, 1993.

Felder, Richard M, Donald R Woods, James E Stice, and Armando Rugarcia. "The Future of Engineering Education: Part 2. Teaching Methods That Work." *Chemical Engineering Education* 34, no. 1 (2000): 26–39.

Forty, Adrian. *Objects of Desire: Design and Society from Wedgwood to IBM*. New York: Pantheon Books, 1986.

Foucault, Michel. "Governmentality." In *The Foucault Effect: Studies in Governmentality*, edited by Graham Burchell, Colin Gordon, and Peter Miller, 87–104. Chicago: University of Chicago Press, [1978]1991.

———. "Politics and the Study of Discourse." In *The Foucault Effect: Studies in Governmentality*, edited by Graham Burchell, Colin Gordon and Peter Miller, 53–72. Chicago: University of Chicago Press, [1968]1991.

Fox, Richard W., and T.J. Jackson, eds. *The Culture of Consumption*. New York: Pantheon, 1983.

Gershinowitz, Harold. "Applied Research for the Public Good-A Suggestion." *Science*, 28 April 1972, 380–86.

Goodchild, Fiona M. "The Pipeline: Still Leaking." *American Scientist*, March-April 2004, 122–14.

Gordon, Colin. "Governmental Rationality: An Introduction." In *The Foucault Effect: Studies in Governmentality*, edited by Graham Burchell, Colin Gordon and Peter Miller, 1–52. Chicago: University of Chicago Press, 1991.

Greenberg, Daniel. "So Many PhDs." *The Washington Post*, July 2 1995, C7.

Handelsman, Jo, Diane Ebert-May, Robert Beichner, Peter Bruns, Amy Chang, Robert DeHaan, Jim Gentile, Sarah Lauffer, James Stewart, Shirley M. Tilghman, and William B. Wood. "Scientific teaching." *Science*, 23 April 2004, 521–22.

Hart, Jeffrey A. *Rival Capitalists: International Competitiveness in the United States, Japan and Western Europe*. Ithaca: Cornell University Press, 1992.

Hawkins, B. Denise. "Colleges must stimulate K-12 pipeline, researchers, policy-makers warn." *Black Issues in Higher Education*, Jan 28 1993, 46–7.

Heilbroner, Robert L. "Priorities for the Seventies." *Saturday Review*, 3 Jan 1970, 17–19.

Hogg, Russell E. "Education and Our National Economic Future." *Vital Speeches of the Day*, no. 15 November (1983): 81–84.

Holden, Constance. "Wanted: 675,000 Future Scientists and Engineers." *Science*, 30 June 1989, 1536–7.

House, Peter, and Roger Shull. *The Practice of Policy Analysis: Forty Years of Art and Technology*. Washington, D.C., 1991.

Howe, Harold. "Education Moves to Center Stage: An Overview of Recent Studies." *Phi Delta Kappan* (1983).

Jackson, Shirley Ann. "The Quiet Crisis: Falling Short in Producing American Scientific and Technical Talent." San Diego, CA: Building Engineering and Scientific Talent (BEST), 2002.

Jackson, Tim. *The Next Battleground: Japan, America, and the New European Market*. New York: Houghton Mifflin Company, 1993.

Jennings, John F. "The Sputnik of the Eighties." *Phi Delta Kappan*, no. October (1987): 104–09.

Jordan, John. *Machine-Age Ideology: Social Engineering and American Liberalism, 1911–1939*. Chapel Hill: University of North Carolina Press, 1994.

Josephson, Paul. "Rockets, Reactors, and Soviet Culture." In *Science and the Soviet Social Order*, edited by Loren R.Graham. Cambridge: Harvard University Press, 1990.

Killian, James. *Sputnik, scientists, and Eisenhower: A memoir of the first special assistant to the President for science and technology*. Cambridge, MA: MIT Press, 1977.

Kinmonth, Earl H. "Japanese Engineers and American Myth Makers." *Pacific Affairs* 64, no. 3 (1991): 328–50.

Kleinman, Daniel. *Politics on the Endless Frontier: Postwar Research Policy in the United States*. Durham, NC: Duke University Press, 1995.

Kreidler, Robert N. "The President's Science Advisers and National Science Policy." In *Scientists and National Policy-Making*, edited by Robert Gilpin and Christopher Wright. New York: Columbia University Press, 1964.

Krieghbaum, Hillier, and Hugh Rawson. *An Investment in Knowledge: The First Dozen Years of NSF's Summer Institutes Programs to Improve Secondary School Science and Mathematics Teaching 1954–1965*. New York: NYU Press, 1969.

Labate, J. "Snapshot of the Pacific Rim." *Fortune*, 7 October 1991, 128–30.

Lepkowski, Wil. "National Science Foundation Embarks on Conservative Course." *C&EN* 1983.

Lessing, Lawrence. "Senseless War on Science." *Fortune*, March 1971, 88–91, 153–55.

Leuenberger, Theodor, and Martin E. Weisntein, eds. *Europe, Japan and America in the 1990s: Cooperation and Competition*. Berlin: Springer-Verlag, 1992.

Lora, Ronald. "Education: Schools as Crucible in Cold War America." In *Reshaping America: Society and Institutions 1945–1960*, edited by Robert H. Bremner and Gary W. Reichard. Columbus: Ohio State University Press, 1982.

Lorriman, John , and Takashi Kenjo. *Japan's Winning Margins: Management. Training and Education*. Oxford: Oxford University Press, 1994.

Lucena, Juan. "Flexible Engineers: History, challenges, and opportunities for engineering education." *Bulletin of Science, Technology, and Society* 23, no. 6 (2003): 419–35.

MacDougall, Walter. *The Heavens and the Earth: A Political Economy of the Space Age*. New York: Basic books, 1985.

Magner, Denise. "Too Many Science Ph.D's?" *The Chronicle of Higher Education*, March 15 1996, A19.

Malcom, Shirley. "Women/Minorities in Science and Technology." *Science* 1981, 137.

Malcom, Shirley, Paula Hall, and Janet W. Brown. *The Double Bind: The Price of Being A Minority Woman in Science*. Washington, D.C.: AAAS, 1976.

Marcus, George. *Ethnography through Thick and Thin*. New York: Princeton University Press, 1999.

McElroy, W. D. "NSF: A Look Ahead." *Science*, January 28 1972, 1.

Merkel, Angela. "The Role of Science in Sustainable Development." *Science*, July 17 1998, 336–7.

Milbank, Dana. "Shortage of Scientists Approaches a Crisis As More Students Drop Out of the Field." *Wall Street Journal*, no. 17 September (1990).

Miller, Peter, and Ted O'Leary. "Accounting and the construction of the governable person." *Accounting, Organizations and Society* 12, no. 3 (1987): 235–65.

Miller, Peter, and Nikolas Rose. "Governing economic life." In *Foucault's New Domains*, edited by Mike Gane and Terry Johnson, 75–105. London: Routledge, 1993.

Montgomery, Scott L. *Minds for the Making: The Role of Science in American Education, 1750–1990*. New York: Guilford Press, 1994.

Morishima, Michio. *Why Has Japan "Succeeded"?: Western Technology and the Japanese Ethos*. Cambridge: Cambridge University Press, 1984.
National Commission on Excellence in Education. "A Nation At Risk: The Imperative for Educational Reform." Washington, D.C.: National Commission on Excellence in Education, 1983.
National Education Association and American Council on Education. "Education and National Security." Washington, D.C.: NEA, 1951.
National Engineering Action Conference. "Agenda and Papers." Paper presented at the National Engineering Action Conference, New York 1982.
National Research Council. *Engineering Education and Practice in the U.S.* Washington, D.C: NAS Press, 1985.
———. *Engineering Infrastructure Diagramming and Modeling*. Washington, D.C.: NAS Press, 1986.
———. *Manpower for Environmental Pollution Control*. Washington, D.C.: NAS Press, 1977.
———. *A Manpower Policy for Primary Health Care*. Washington, D.C.: NAS Press, 1978.
National Research Council, Committee on Science and Technology for Countering Terrorism. "Making the Nation Safer: The role of science and technology in countering terrorism." 23. Washington, D.C.: The National Academies, 2002.
National Research Council, Office of Scientific and Engineering Personnel. "Forecasting Demand and Supply of Doctoral Scientists and Engineers: Report of a Workshop on Methodology." Washington, D.C: NAS, 2000.
National Science and Technology Council. "Ensuring a Strong U.S. Scientific, Technical, and Engineering Workforce in the 21st Century." Washington, D.C: OSTP, 2000.
National Science Board. *The Role of the National Science Foundation in Economic Competitiveness*. Washington, D.C.: NSF, 1988.
———. *Science and Engineering Indicators 1987*. Washington, D.C.: NSF, 1987.
———. *Science and Engineering Indicators 2004*. Arlington: NSF, 2004.
———. "The Science and Engineering Workforce: Realizing America's Potential." Arlington, VA: NSF, 2003.
———. *Scientific and Technical Manpower Projections, Report of the ad hoc Committee on Manpower*. Washington, D.C.: NSF, 1974.
National Science Foundation. *Annual Report*. Washington, D.C.: NSF, 1962.
———. *Annual Report*. Washington, D.C.: NSF, 1956.
———. *Annual Report*. Washington, D.C.: NSF, 1958.
———. *Annual Report*. Washington, D.C.: NSF, 1959.
———. *Annual Report*. Washington, D.C.: NSF, 1960.
———. *Annual Report*. Washington, D.C.: NSF, 1961a.
———. *Annual Report*. Washington, D.C.: NSF, 1964a.
———. *Annual Report*. Washington, D.C.: NSF, 1970.
———. *Annual Report*. Washington, D.C., 1974.
———. *Annual Report*. Washington, D.C.: NSF, 1975.
———. *Annual Report*. Washington, D.C.: NSF, 1988.

——. *Annual Report*. Washington D.C.: NSF, 1957.

——. *Annual Report*. Washington, D.C: NSF, 1989.

——. "Engineering Education Coalitions." Washington, D.C.: NSF, 1993.

——. *Engineering: A More Expanded Role for NSF*. Washington, D.C.: NSF, 1985.

——. *Enhancing the Quality of Education and Human Resources for Science, Mathematics, and Engineering in the United States: Strategic Plan for NSF FY 1991–FY 1995*. Washington, D.C.: NSF, 1991.

——. "Grant Proposal Guide." Arlington, VA: NSF, 2003.

——. *Indicators of Science and Mathematics Education 1992*. Washington, D.C.: NSF, 1993b.

——. *The Learning Curve: What We Are Discovering About U.S. Science and Mathematics Education*. Washington, D.C.: NSF, 1996.

——. *Long Range Plan FY1989–1993*, 1988.

——. *The Long-Range Demand for Scientific and Technical Personnel: A Methodological Study*. Washington, D.C.: NSF, 1961b.

——. "NSF in a Changing World." Washington, D.C.: NSF, 1995.

——. *NSF Overseas Offices* [Webpage]. September 2004 [cited. Available from http://www.nsf.gov/home/int/ovoff.htm.

——. *NSF's Programmatic Reform: The Catalyst for Systemic Change*. Washington, D.C.: NSF, 1994.

——. "Restructuring Engineering Education: A Focus on Change." Arlington, VA: NSF, 1995.

——. *Scientific and Technical Manpower Resources: Summary Information on Employment, Characteristics, Supply and Training*. Washington, D.C.: NSF, 1964b.

——. *Scientists, Engineers, and Technicians in the 1960's: Requirements and Supply*. Washington, D.C.: NSF, 1963.

——. *The State of Academic Science and Engineering*. Washington, D.C.: NSF, 1990.

——. *Women, Minorities, and Persons with Disabilities in Science and Engineering*. Arlington, VA: NSF, 2004.

National Science Foundation, Advisory Committee for Science Education. "Science Education -The Task Ahead for the NSF." Washington, D.C.: NSF, 1970.

National Science Foundation, Directorate for Education and Human Resources. "Decade of Achievement: Educational Leadership in Mathematics, Science and Engineering." Arlington, VA: NSF, 1992.

National Science Foundation, Division of Science Resources Statistics. "SESTAT and NIOEM: Two Federal Databases Provide Complementary Information on the Science and Technology Labor Force." Arlington, VA.: National Science Foundation, 1999.

National Science Foundation, Division of Science Resources Studies. "Graduate Education Reform in Europe, Asia, and the Americas and International Mobility of Scientists and Engineers: Proceedings of an NSF Workshop." Arlington, VA: NSF, 2000.

National Science Foundation, Policy Research Analysis Division. "Demand-Supply Balance for Scientists and Engineers Within Government, Industry and Academic Sectors." Washington, D.C.: NSF, 1984.

———. "Personnel in Natural Science and Engineering: Working Draft." Washington, D.C.: NSF, 1988.

Parks, Lisa. "Technology in the Twilight: A Cultural History of the Communications Satellite 1955–58." Paper presented at the 1995 Society for the History of Technology Conference,, Charlottesville, Virginia, Oct 18–22 1995.

Peden, Irene C., Edward W. Ernst, and John W. Prados. "Systemic Engineering Education Reform: An Action Agenda. Recommendations of a Workshop Convened by the NSF Engineering Directorate." Washington, D.C.: NSF, 1995.

Prados, John. "Engineering Criteria 2000–A Change Agent for Engineering Education." *Journal of Engineering Education* (1997): 69–70.

President's Commission on Industrial Competitiveness. "Global Competition: The New Reality." Washington, D.C., 1985.

President's Science Advisory Committee. *Meeting Manpower Needs in Science and Technology*. Washington, D.C., 1963.

———. *Meeting Manpower Needs in Science and Technology. Report Number One: Graduate Training in Engineering, Mathematics, and Physical Sciences*. Washington, D.C., 1962.

Rickover, Hyman G. *Education and Freedom*. New York, 1959.

Riley, Richard W. "The Goals 2000: Educate America Act. Providing a World-Class Education for Every Child." In *National Issues in Education. Goals 2000 and School-to-Work*, edited by John F. Jennings. Bloomington, IN: Phi Delta Kappa, 1995.

Rose, Nikolas. *Governing the Soul: The Shaping of the Private Self*. London: Routledge, 1990.

Rosenblatt, Alfred. "Who's Ahead in Hi-Tech." *IEEE Spectrum* 28, no. 4 (1991): 22–27.

Schmitt, Roland W. "Engineering, Engineers, and Engineering Education in the Twenty-First Century." Belmont, Maryland: National Science Foundation and National Academy of Engineering, 1990.

Seymour, Elaine, and Nancy Hewitt. *Talking About Leaving: Why Undergraduates Leave the Sciences*. Boulder, CO: Westview Press, 2000.

Shaw, Stephen. "The Coming Crisis in Aerospace Employment." *Aviation Week* 1990, 32–42.

Smaglik, Paul. "Patching a leaky pipeline." *Nature*, February 12 2004, 657.

Smith, Charles. "Sputnik II-Where Are You When We Need You." *Change*, October 1983, 7–11.

Smith, Marshall S., and Jennifer O'Day. "Systemic School Reform." In *The Politics of Curriculum and Testing*, edited by Susan H. Fuhrman and Betty Malen. Philadelphia: Falmer, 1991.

Spigel, Lynn. "From Domestic Space to Outer Space." In *Close Encounters: Film, Feminism and Science Fiction*, edited by et al. Constance Penley. Minneapolis: University of Minneapolis Press, 1991.

Springer, L., M.E. Stanne, and S. S. Donovan. "Effects of small-group learning on undergraduates in science, mathematics, engineering, and technology: A meta-analysis." *Review of Educational Research* 69, no. 1 (1999): 21–51.

Stockton, William. "The Technology Race." *New York Times Magazine*, 28 June 1981, 14–55.

Task Force on Women, Minorities and the Handicapped in Science and Technology. "Changing America: The New Face of Science and Engineering." Washington, D.C., 1989.

Teitelbaum, Michael S. "Do we need more scientists?" *The Public Interest*, no. 153 (2003): 40–53.

Thimann, Kenneth. "Science: Attack and Defense." *Science*, August 14 1970, 1.

Thomas, Thomas G., and Mohammad S. Alam. "Addressing ABET 2000 Requirements for Continual Evaluation and Program Improvement." Paper presented at the 2003 ASEE Annual Conference, Nashville, TN 2003.

Tully, Shawn, and Erik Schonfeld. "Europe 1992: More Unity than You Think." *Fortune*, 24 August 1992, 136–41.

U.S. Congress Joint Committee on Atomic Energy. *Development of Scientific, Engineering, and Other Professional Manpower*, 85th Cong., 1st sess., 1957.

———. *Engineering and Scientific Manpower in the U.S., Western Europe and Soviet Republics*, 84th Cong., 2nd sess., 1956.

U.S. Congress Joint Economic Committee. *Investment in Research and Development*, 100th Cong., 1st sess., 1987.

U.S. Department of Education. "Japanese Education Today." Washington, D.C., 1987.

U.S. Department of Health, Education, and Welfare. "Science as a Way of Life." 1961.

U.S. House Committee on Appropriations. *HUD and Certain Independent Agencies Appropriations, FY1977*, 94th Cong., 2nd sess., 1976.

———. *National Science Foundation: Comparison of United States and USSR Science Education.*, 86th Cong., 2nd sess., 1960.

U.S. House Committee on Education and Labor. *Science and Mathematics Education Improvement Act*, 97th Cong., 2nd sess., 1982.

U.S. House Committee on Education and Labor, Special Subcommittee on Education. *National Defense Education Act*, 87th Cong., 1st sess., 1961.

U.S. House Committee on Science and Astronautics. *1972 NSF Authorization*, 92nd Cong., 1st sess., 1971.

———. *1973 NSF Authorization*, 92nd Cong., 2nd sess., 1972.

———. *The National Science Foundation: Its Present and Its Future*, 89th Cong., 2nd sess., 1966.

———. *Science Manpower and Education*, 86th Cong., 1st sess., 1959.

———. *Scientific and Technical Manpower, Supply, Demand, and Utilization.*, 87th Cong., 2nd sess., 1962.

———. *A Study of Scientific and Technical Manpower: A Program of Collection, Tabulation, and Analysis of Data of the National Science Foundation*, 86th Cong., 2nd sess., 1960.

U.S. House Committee on Science and Technology. *Engineering Manpower Concerns*, 97th Cong., 1st sess., 1981.

———. *Forecasting Needs for the High Tech Industry*, 97th Cong., 1st sess., 1981.

———. *H.R.6910 National Technology Foundation Act of 1980*, 96th Cong., 2nd sess., 1980.

———. *H.R. 5254: Engineering and Science Manpower Act of 1982*, 97th Cong., 2nd sess., 1982.

———. *NSF 1976 Posture Hearing*, 94th Cong., 1st sess., 1975.

———. *Oversight Hearings on NSF Science Education Programs*, 94th Cong., 1st sess., 1975.

———. *Science and Engineering Education and Manpower*, 97th Cong., 2nd sess., 1982.

———. *U.S. Science and Technology Under Budget Stress*, 97th Cong., 1st sess., 1981.

U.S. House Committee on Science and Technology, Task Force on Science Policy. *Science Policy Study? Hearings Vol 9: Scientists and Engineers: Supply and Demand*, 99th Cong., 1st sess., 1985.

U.S. House Committee on Science, Space and Technology. *Projecting Science and Engineering Personnel Requirements for the 1990s: How Good Are the Numbers?*, 103rd Cong., 1st sess., 1993.

U.S. House Committee on Science, Space, and Technology, Subcommittee on Science. *The Mission of the National Science Foundation*, 103rd Cong., 1st sess., 1993.

U.S. Office of Education. "Life Adjustment for Every Youth." Washington, D.C., 1951.

U.S. Senate Committee on Governmental Affairs. *Department of Education Organization Act of 1979*, 96th Cong., 1st sess., 1979.

U.S. Senate Committee on Labor and Human Resources. *NSF Authorization Act for FY1983*, 97th Cong., 2nd sess., 1982.

U.S. Senate Committee on Labor and Public Welfare. *National Defense Education Act of 1958 - Summary and Analysis*, 85th Cong., 2nd sess., 1958.

———. *NSF Authorization Legislation, 1976*, 94th Cong., 2nd sess., 1976.

———. *NSF Authorization, 1971*, 91st Cong., 2nd sess., 1970.

———. *NSF Authorization, 1972*, 92nd Cong., 1st sess., 1971.

———. *NSF Authorization, 1973*, 93rd Cong., 1st sess., 1972.

———. *NSF Legislation, 1973*, 93rd Cong., 1st sess., 1973.

———. *NSF Legislation, 1974*, 93rd Cong., 2nd sess., 1974.

———. *Science and Education for National Defense*, 85th Cong., 2nd sess., 1958.

Vaughn, John C. "Heading Off a Ph.D. Shortage." *Issues in Science and Technology*, Winter 1990, 66–72.

Vetter, Betty. "The Technological Marketplace: Supply and Demand for Scientists and Engineers." Washington, D.C.: Scientific Manpower Commission, 1985.

———. "Women and Minority Scientists." *Science*, September 5 1975, 751.

———. "Women, Men and the Doctorate." *Science*, January 31 1975, 301.

Walsh, John. "Knapp Reinterprets Excellence at NSF." *Science*, 2 December 1983, 990–2.

Whitfield, Stephen J. *The Culture of the Cold War*. Baltimore: John Hopkins, 1991.

Winner, Langdon. "Upon Opening the Black Box and Finding it Empty: Social Constructivism and the Philosophy of Technology." *Science, Technology and Human Values* 1991, 503–19.

# Index